Minerals, Rocks and Inorganic Materials

Monograph Series of Theoretical and Experimental Studies

5

Edited by

W. von Engelhardt, Tübingen · T. Hahn, Aachen
R. Roy, University Park, Pa. · P. J. Wyllie, Chicago, Ill.

Subseries:
Isotopes in Geology

G. Faure · J. L. Powell

Strontium Isotope Geology

With 51 Figures

Springer-Verlag Berlin · Heidelberg · New York 1972

Professor *Gunter Faure*
Department of Geology
The Ohio State University
Columbus, Ohio/USA

Professor *James L. Powell*
Department of Geology
Oberlin College
Oberlin, Ohio/USA

ISBN 3-540-05784-6 Springer-Verlag Berlin Heidelberg New York
ISBN 0-387-05784-6 Springer-Verlag New York Heidelberg Berlin

Preface

Since the end of World War II isotope geology has grown into a diversified and complex discipline in the earth sciences. It has progressed by the efforts of a relatively small number of specialists, many of whom are physicists, chemists, or mathematicians who were attracted to the earth sciences by the opportunity to measure and to interpret the isotopic compositions of certain chemical elements in geological materials. The phenomenal growth of isotope geology during the last 25 years is an impressive indication of the success of their efforts.

We have now entered into a new phase of development of isotope geology which emphasizes the application of the new tools to the solution of specific problems in the earth and planetary sciences. This requires the active participation of a new breed of geologists who understand the nature and complexity of geological problems and can work toward their solution by a thoughtful application of the principles of isotope geology. It is therefore necessary to explain these principles to earth scientists at large to enable them to make use of the new information which isotope geology can offer them.

This book was written to provide this kind of an introduction to the isotope geology of strontium. The isotopic composition of this chemical element, which is widely dispersed in most terrestrial and extraterrestrial rocks, is changed continuously as a result of the radioactive decay of naturally-occurring ^{87}Rb to stable ^{87}Sr. The resulting changes in the isotopic abundance of ^{87}Sr can be used to date rocks and to obtain information about their origin. We have tried to explain the theoretical basis for these interpretations in such a way that they can be understood by nonspecialists in the earth sciences. Although this subject is relatively young, its literature is voluminous and widely scattered. We have aimed for comprehensive explanations of ideas and concepts, but make no claim to an exhaustive review of the literature.

In the early 1960's attempts to understand the distribution of strontium isotopes in nature were hampered by poor analytical precision and the resulting interlaboratory discrepancies. However, by about 1964 most laboratories were reporting results for an interlaboratory strontium-isotope standard that agreed within about $\pm 0.15\%$. In this book the geologic conclusions drawn from strontium isotope data are generally based on many analyses made at several different laboratories; they do not depend on small differences in ${}^{87}Sr$ abundance for a small number of analyzed samples. Therefore, the data employed here generally have not been corrected to a common value for an interlaboratory strontium-isotope standard.

We dedicate this book to the memory of ARTHUR HOLMES, who was one of the first earth scientists to appreciate the importance of radioactivity and the resulting variation of isotopic compositions to geology. Many applications of isotopic studies to the solution of geological problems which he foresaw are now becoming a reality. This book contains only a small number of the results of scientific research in isotope geology which he stimulated.

We have benefited from helpful comments and discussions of several of our colleagues. We are especially grateful to ROBERT J. PANKHURST, who read the entire manuscript and helped us to improve it. We were also helped by PATRICK M. HURLEY, NORMAN K. GRANT, GERALD J. WASSERBURG, and WARREN HAMILTON, who read specific chapters and offered constructive comments. However, we take sole responsibility for any errors or omissions. This book is a joint venture, and while our names appear in alphabetical order, we share equally in the responsibility for its contents.

It is a pleasure to acknowledge the friendly relationship which we have enjoyed with our patient editor, Dr. PETER J. WYLLIE. We set and passed many deadlines without once being taken to task for the delays in completing the manuscript.

We thank our efficient secretaries who typed this manuscript: At Ohio State University we thank Miss CARLA MORRIS and Mrs. HELEN JONES, and at Oberlin College we acknowledge Mrs. JEANNE BOWMAN and Mr. HUBERT BATES.

Above all, we hope that this book will be useful to advanced undergraduate and graduate students in geology departments everywhere.

GUNTER FAURE
JAMES L. POWELL

Contents

I. The Geochemistry of Rubidium and Strontium

1. Introduction

Most igneous, metamorphic, and sedimentary rocks contain rubidium (Rb) and strontium (Sr) in detectable amounts. However, the concentrations of these elements are almost always less than 1 percent, and they are therefore rarely determined in routine chemical analyses. Neither rubidium nor strontium is a major constituent in the common rock-forming silicate minerals, although strontium does form a carbonate (strontianite) and a sulfate (celestite) which are found in some hydrothermal deposits and certain sedimentary rocks, particularly carbonates.

The importance of these two elements to the earth sciences lies in two facts. First, one of the naturally-occurring isotopes of rubidium (^{87}Rb) is radioactive and decays to a stable isotope of strontium (^{87}Sr). Therefore, the amount of ^{87}Sr in a mineral or rock containing rubidium increases continuously as a function of time. This phenomenon is used to make age determinations of rocks and minerals by the Rb-Sr method. Secondly, radiogenic ^{87}Sr can be treated as a geological "tracer" to study certain geological processes which are of interest to petrologists. In fact, studies of the isotopic composition of strontium have contributed significantly to our knowledge of the petrogenesis of igneous rocks and to our understanding of the chemical evolution of the Earth.

The objective of this book is to explain the concepts which underlie the interpretations of the isotopic composition of strontium in rocks and to present the important conclusions which have been derived. This chapter contains brief summaries of the occurrence and the distribution of rubidium and strontium in minerals and rocks and of the factors controlling them. More detailed descriptions of the geochemistry of rubidium and strontium can be found in geochemistry books of RANKAMA and SAHAMA (1950), GOLDSCHMIDT (1954), MASON (1958), and VLASOV (1964a, b).

2. Rubidium

Rubidium is a member of Group IA of the Periodic Table, which includes hydrogen and the alkali metals: lithium, sodium, potassium, rubidium, cesium, and francium. All of the alkali metals have a single valence electron in an s orbital outside of stable electronic configurations. This valence electron is readily removed to form ions having a charge of $+1$. The alkali metals have low electronegativities and therefore form strongly ionic bonds with nonmetallic elements, such as oxygen or the halogens. Their ionic radii increase from 0.60 Å for Li^+ to 1.69 Å for Cs^+ and are large compared to those of other metals. The physical and chemical parameters of the alkali metals which we have discussed are summarized in Table I.1.

Table I.1. Physical and chemical properties of the alkali metals

Properties	Li	Na	K	Rb	Cs
Atomic number	3	11	19	37	55
Atomic weight (based on C^{12})	6.939	22.9898	39.102	85.467	132.905
Ionic radius (Å) (PAULING)	0.60	0.95	1.33	1.48	1.69
Radius ratio ($O^{2-} = 1.40$ Å)	0.43	0.68	0.95	1.06	1.21
Coordination number in ionic crystals	6	6,8	8,12	8,12	12
Electronegativity (PAULING)	1.0	0.9	0.8	0.8	0.7
Percent ionic character of bond with O^{2-}	82	83	87	87	89

The radius ratio (radius of the cation divided by the radius of anion in an ionic crystal) is a useful indicator of the preferred coordination number of cations under conditions of closest packing of ions. Data from KRAUSKOPF (1967) and COTTON and WILKINSON (1962).

Rubidium was discovered in 1861 by BUNSEN and KIRCHHOFF in the mineral lepidolite. The distribution of this element and its congeners in rocks and minerals has been studied by GOLDSCHMIDT et al. (1933), GOLD-SCHMIDT et al. (1934), AHRENS, PINSON, and KEARNS (1952), HORSTMAN (1957), TAYLOR and HEIER (1958), HEIER and TAYLOR (1959), and many other geochemists. The most comprehensive treatment of the geochemistry of the alkali metals is by HEIER and ADAMS (1964), who also gave a very complete listing of references to the literature.

The distribution of rubidium in nature is governed primarily by the fact that the Rb^+ ion ($r = 1.48$ Å) is small enough to be admitted into K^+ sites ($r = 1.33$ Å) in all of the important rock-forming minerals that contain potassium, whereas Cs^+ ($r = 1.69$ Å) is so large that it is concentrated in the fluid phase and eventually may even form cesium minerals such as pollucite ($Cs[AlSi_2O_6]$). Rubidium is never concentrated sufficiently to form its own minerals, but is widely dispersed as a trace element in potassium-bearing minerals.

The principal carriers of rubidium in igneous and metamorphic rocks are the micas (biotite, muscovite, and lepidolite) and the potassium feldspars (orthoclase and microcline). Minerals in pegmatites may contain appreciably higher concentrations of rubidium than the same minerals occurring in ordinary igneous or metamorphic rocks. Lepidolite from lithium-bearing pegmatites may contain several percent of rubidium. The concentration of rubidium in plagioclase feldspar is low because the Rb^+ ion is too large to replace the Na^+ ion, which has a radius of only 0.95 Å. Rubidium also occurs in other rock-forming silicate minerals such as the pyroxenes and amphiboles. However, its concentration in these minerals is commonly less than 10 parts per million (ppm) by weight.

The apparent close geochemical coherence between potassium and rubidium has led to the suggestion that the K/Rb ratio of igneous rocks is constant [AHRENS, PINSON, and KEARNS (1952)]. AHRENS and TAYLOR (1960) recommended an average K/Rb ratio of 230 for igneous rocks. However, more recent studies reviewed by TAUBENECK (1965) and TAYLOR (1965), show that the K/Rb ratio decreases during progressive differentiation of granitic magmas because igneous processes generally concentrate rubidium in the residual magma to a greater extent than potassium. Thus the K/Rb ratio of suites of comagmatic granitic rocks tends to decrease with increasing potassium content. The enrichment of rubidium relative to potassium in the residual liquid of a magma undergoing crystallization is due to the difference in ionic radii of K^+ and Rb^+ ions. Since Rb^+ is larger than K^+, its charge-to-radius ratio is less, and it is therefore less strongly attracted by the negative charges surrounding K^+ sites in the crystal lattices of rock-forming silicate minerals. Table I.2 is a summary of K/Rb ratios in igneous rocks taken from a review by ERLANK (1968).

Table I.2. Typical K/Rb ratios in igneous rocks

Type of igneous rock	Ratio
Granites, granodiorites, etc.	50 — 350
Rhyolites, etc.	100 — 350
Syenites	250 — 700
Basalt (tholeiitic, continental)	150 —1000
Basalt (tholeiitic, oceanic)	450 —2000
Ultramafic rocks (alpine type, continental)	200 — 400

After ERLANK (1968).

Rubidium has two naturally-occurring isotopes: ^{85}Rb and ^{87}Rb. In addition, a large number (about 20) of short-lived radioactive isotopes of rubidium have been produced by nuclear reactions under laboratory conditions. The abundance ratio of ^{85}Rb to ^{87}Rb of terrestrial rubidium was recently redetermined by CATANZARO et al. (1969). They reported that the ratio $^{85}Rb/^{87}Rb$ is 2.59265 ± 0.00170 and that the corresponding isotopic abundances are ^{85}Rb = 72.1654 ± 0.0132 percent and ^{87}Rb = 27.8346 ± 0.0132 percent. The atomic weight (on the ^{12}C scale) of rubidium is 85.46776

$\pm$ 0.00026. Earlier, SHIELDS et al. (1963), had determined the isotopic composition of rubidium extracted from 27 silicate minerals ranging in age from 20 to 2600 million years. They found that the $^{85}Rb/^{87}Rb$ ratio in all of the samples is constant. This result confirms the widely-held view that all naturally-occurring rubidium, regardless of the age of the mineral in which it is found or of its geochemical history, has the *same* isotopic composition. The apparent uniformity of the isotopic composition of terrestrial rubidium, and of many other elements as well, suggests that the elements were thoroughly mixed in the solar dust cloud before being incorporated into the earth.

3. Strontium

Strontium is a member of Group II A of the Periodic Table, which includes the alkaline earths: beryllium, magnesium, calcium, strontium, barium, and radium. All of the alkaline earth elements have two valence electrons in an *s* orbital outside of stable noble gas electron configurations and readily form ions having a charge of $+2$. Their electronegativities are low, ranging from 1.5 for beryllium to 0.9 for barium on Pauling's scale. Consequently, the alkaline earths form ionic bonds with nonmetallic elements, including oxygen. Table I.3 lists some of the important physical and chemical parameters of the elements in this group.

Table I.3. Physical and chemical properties of the alkaline earths

Properties	Be	Mg	Ca	Sr	Ba
Atomic number	4	12	20	38	56
Atomic weight (based on ^{12}C)	9.012	24.312	40.08	87.62	137.34
Ionic radius (Å) (PAULING)	0.31	0.65	0.99	1.13	1.35
Radius ratio (O^{2-} = 1.40 Å)	0.22	0.46	0.71	0.81	0.96
Coordination number in ionic crystals	4	6	6, 8	8	8, 12
Electronegativity (PAULING)	1.5	1.2	1.0	1.0	0.9
Percent ionic character of bond with O^{2-}	63	71	79	82	84

Data from KRAUSKOPF (1967), and COTTON and WILKINSON (1962).

The distribution of strontium in nature and its geochemical properties have been studied by NOLL (1934), HEVESY and WÜRSTLIN (1934), TUREKIAN and KULP (1956), ODUM (1957), and TUREKIAN (1964). The most recent summary of the geochemistry of strontium appears to be by TAYLOR (1965).

The distribution of strontium in rocks is controlled by the extent to which Sr^{2+} can substitute for Ca^{2+} in calcium-bearing minerals and the degree to which potassium feldspar can capture Sr^{2+} in place of K^+ ions.

Although the ionic radius of Sr^{2+} (1.13 Å) is only about 15 percent greater than that of Ca^{2+} (0.99 Å), its radius ratio relative to O^{-2} is 0.81, while that of Ca^{2+} is 0.71. This means that Sr^{2+} favors eight-fold coordination while Ca^{2+} is able to occupy both six- and eight-fold coordinated lattice positions. As a result, strontium acts as a dispersed trace element in igneous rocks, but can be concentrated relative to calcium to enable it to form its own minerals in hydrothermal deposits and carbonate rocks. In fact, VLASOV (1964a), lists 27 strontium minerals, of which only two (celestite and strontianite) are important.

The principal carriers of strontium in igneous rocks are plagioclase feldspar and apatite, in which Sr^{2+} can replace Ca^{2+} ions. The strontium content of pyroxenes is generally low because calcium is bonded to six oxygen atoms, and its lattice site is therefore too small for Sr^{2+} ions. The substitution of Ca^{2+} by Sr^{2+} in plagioclase and pyroxene has been discussed by BROOKS (1968). Potassium feldspar can capture Sr^{2+} ions in place of K^+ ions. The substitution is presumably coupled with the replacement of Si^{4+} by Al^{3+} in the silica tetrahedra in order to maintain electrical neutrality. However, the capture of Sr^{2+} for K^+ in micas is not favored because potassium in micas has 12-fold coordination, making this site too large for strontium. The strontium content of some biotites may be partly due to the presence of inclusions of apatite.

During crystallization of magma, strontium initially enters calcic plagioclase by substitution for Ca^{2+}. If differentiation progresses to the point that potassium-feldspar begins to form, Sr^{2+} ions are captured from the residual liquid by K^+ sites. Consequently, in many series of differentiated igneous rocks, such as those studied by NOCKOLDS and MITCHELL (1948), and by NOCKOLDS and ALLEN (1953, 1954, and 1956), the strontium concentration *decreases* with increasing degree of fractionation of the magma. If pyroxene or olivine form before plagioclase, the strontium concentration of the residual liquid may rise to a maximum before declining.

Strontium has four stable isotopes: ^{88}Sr, ^{87}Sr, ^{86}Sr, and ^{84}Sr. In addition, 14 short-lived radioactive isotopes of strontium have been made artifically. Among these is ^{90}Sr, which is a product of nuclear fission of uranium. It is considered to be very dangerous because it is concentrated in bones when ingested and because the beta particles and gamma rays which it emits destroy tissue.

The isotopic composition of strontium in nature is not constant, but depends on the Rb/Sr ratio of the sample from which the strontium is extracted and on the length of time it has been associated with rubidium. The relative abundance of ^{87}Sr among the isotopes of strontium is commonly expressed as the atomic ratio $^{87}Sr/^{86}Sr$. NIER (1938) reported the following abundance ratios for purified strontium metal: $^{87}Sr/^{86}Sr = 0.7119$; $^{86}Sr/^{88}Sr = 0.1194$; $^{84}Sr/^{88}Sr = 0.0068$. His value for the $^{86}Sr/^{88}Sr$ ratio is still used as a reference to eliminate the effect of isotopic fractionation from measured values of the $^{87}Sr/^{86}Sr$ ratio in geological materials.

Several geochemists have attempted to estimate the average concentrations of the major and trace elements in different rock types and in the crust of the Earth. Such compilations have been published by TAYLOR (1964), TUREKIAN and WEDEPOHL (1961), GREEN (1959), and VINOGRADOV (1956, 1962), among others. We have chosen the values compiled from the literature by TAYLOR (1965), to represent the average concentrations of rubidium and strontium in different rock types and in the crust of the Earth. TAYLOR's estimates appear in Table I.4 along with the corresponding Rb/Sr ratios. These averages indicate the general levels of concentrations of rubidium and strontium in rocks, but we emphasize that large variations exist among each of the rock types listed in Table I.4.

Table I 4. Average concentrations of rubidium and strontium in terrestrial rocks and chondritic meteorites (ppm)

	Rb	Sr	Rb/Sr
Chondrites	2.3	10	0.23
Crust	90	375	0.24
Ultrabasic rocks	0.077—7.75	2.32—72.4	0.007—1.32
Basalt	30	465	0.06
Syenite	110	300	0.37
Granodiorite	120	450	0.27
Granite	150	285	0.53
Shale	140	300	0.47
Greywacke	120	450	0.27
Quartzite	30	—	—
Limestone	5	500	0.01

Data from a compilation by TAYLOR (1965), except for the ultrabasic rocks, which are from analyses by STUEBER and MURTHY (1966).

4. The Rubidium/Strontium Ratio

We have discussed in the foregoing sections the fact that rubidium is concentrated in the residual fluid during fractional crystallization of magma and that it eventually enters potassium minerals. Strontium, on the other hand, is removed from the liquid phase and is concentrated primarily in early-formed calcic plagioclase. As a consequence, the Rb/Sr ratio of differentiated igneous rocks tends to increase with increasing degree of differentiation. This phenomenon has been documented by many investigators. To illustrate it we have chosen a fairly typical example from analyses of the trace element composition of the Southern California Batholith reported by NOCKOLDS and ALLEN (1953).

The concentrations of rubidium and strontium (determined by an optical spectrographic technique) are plotted in Fig. I.1 against a parameter which reflects the degree of differentiation of the residual liquid of a magma in the process of crystallization. It can be seen that the concentration of strontium decreases very slightly at first, but more rapidly later with increasing differentiation. The distribution of rubidium is almost the exact opposite to

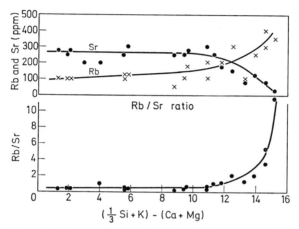

Fig. I.1. Concentrations of rubidium and strontium and the Rb/Sr ratio of granitic rocks from the Southern California Batholith. The graph shows that the rubidium concentration of the more highly differentiated rocks increases as their strontium content decreases. As a result, the Rb/Sr ratio of this and similar granitic rocks ranges from less than 0.5 to values greater than 10. (Data are from NOCKOLDS and ALLEN, 1953)

that of strontium. The rubidium concentration increases very slightly at first, but increases significantly in the more highly-differentiated rocks. The resulting Rb/Sr ratio is nearly constant at first and has an average value somewhat less than 0.5. With increasing degree of differentiation, the Rb/Sr ratio increases steeply to values greater than 10.

This variation of the Rb/Sr ratio of the Southern California Batholith, and many other differentiated igneous rocks, is important for two reasons: (1) The rocks and minerals in which rubidium is concentrated will eventually contain more radiogenic ^{87}Sr produced by decay of ^{87}Rb than rocks or minerals from the same series which are enriched in strontium and depleted in rubidium. By comparing the ^{87}Sr-enrichment of rubidium-rich rocks with that of strontium-rich rocks from the same suite, precise age determinations can be made. (2) In the process of dating, or by analysis of strontium-rich and rubidium-depleted rocks, it is possible to determine the $^{87}Sr/^{86}Sr$ ratio of the magma at the time of its crystallization. The numerical value of this "initial" $^{87}Sr/^{86}Sr$ ratio can be used to infer the source of the magma. These statements will be examined and elaborated in subsequent chapters.

5. Summary

Rubidium and strontium are trace elements which occur in most igneous, metamorphic, and sedimentary rocks in detectable amounts. Rubidium replaces potassium in rock-forming silicate minerals, such as the micas and potassium-feldspar. Strontium can replace calcium, and it occurs in igneous rocks primarily in calcic plagioclase and apatite. Rubidium and strontium are important to the earth sciences because of the radioactive decay of naturally-occurring ^{87}Rb to stable ^{87}Sr a process which can be used for dating rubidium-bearing rocks and minerals and for studying the petrogenesis of igneous rocks.

II. Measurement of Geologic Time by the Rubidium-Strontium Method

1. Introduction

The natural radioactivity of rubidium was first reported by J. J. THOMSON in 1905 and later confirmed by CAMPBELL and WOOD (1906). However, the radioactive isotope of rubidium in nature was not positively identified until 1937. In that year HAHN, STRASSMAN, and WALLING (1937) separated strontium from a specimen of lepidolite from the Silver Leaf Mine in Manitoba, and MATTAUCH (1937) showed by mass spectroscopy that 99.7 percent of this strontium was ^{87}Sr. Since rubidium was known to emit negatively-charged beta particles, it was clear that the ^{87}Sr in this lepidolite had formed by decay of ^{87}Rb. This conclusion was confirmed by HEMMENDINGER and SMYTHE (1937), who demonstrated that ^{87}Rb was the only naturally-occurring radioactive isotope of rubidium.

The application of this phenomenon to the measurement of the ages of old, rubidium-rich minerals was first suggested by GOLDSCHMIDT (1937). HAHN and WALLING (1938) discussed in considerable detail the feasibility and usefulness of the "strontium method" of dating minerals such as lepidolite or potassium feldspar, which may contain several percent rubidium. A few years later, HAHN et al. (1943) published an age determination based on the decay of ^{87}Rb of the mineral pollucite from Varutrask in Sweden, for which they obtained an age of 530 million years.

These early age determinations were applicable only to rubidium-rich minerals and relied on gravimetric determinations of the concentrations of rubidium and strontium, combined with a determination of the isotopic composition of strontium on a mass spectrograph such as the one described by MATTAUCH (1938, 1947). In order to use the "strontium method" to date more common, but less rubidium-rich minerals, such as biotite, muscovite, and potassium feldspar, more sensitive and more accurate analytical techniques were required. These did not become available until about 1950, when research in geochronology was begun in the laboratories of the Carnegie Institution of Washington, D.C., and elsewhere. Since then, analytical techniques and instrumentation have been improved to the point that many minerals, as well as igneous, metamorphic, and certain sedimentary rocks now can be dated reliably by the Rb-Sr method. In this chapter we shall examine the theoretical basis for the measurement of geologic ages using the decay of ^{87}Rb to ^{87}Sr, and lay the groundwork for the interpretation of the isotopic composition of strontium in the rocks and minerals of the earth.

2. Radioactive Decay of ^{87}Rb to ^{87}Sr

^{87}Rb decays to stable ^{87}Sr by beta emission, as follows:

$$^{87}\text{Rb} \rightarrow {}^{87}\text{Sr} + \beta^- + \nu + Q, \tag{II.1}$$

where $\beta^- =$ beta particle having an electronic charge of -1, $\nu =$ neutrino, and $Q =$ decay energy, measured in units of million electron volts (Mev). The value of Q for the decay of ^{87}Rb is 0.275 Mev, which is unusually low for beta-decay processes. The beta particles emitted by the decaying ^{87}Rb nuclei have kinetic energies ranging from essentially 0 up to a maximum of 0.275 Mev. In each decay the kinetic energy of the neutrino is the difference between the maximum decay energy and the kinetic energy of the beta particle. Since neutrinos interact sparingly with matter, they carry their kinetic energy with them into space. The beta particles, on the other hand, are quickly slowed down by collisions with electrons, and their kinetic energy is converted to thermal energy. The low kinetic energy of the beta particles emitted by ^{87}Rb makes their accurate detection difficult and gives rise to a significant uncertainty in the measurement of the specific decay rate of ^{87}Rb.

The rate of decay of a radioactive parent N to a stable daughter D is proportional to the number of parent atoms present. This law of radioactive decay can be expressed as a differential equation:

$$\frac{-dN}{dt} = \lambda N, \tag{II.2}$$

where λ is the decay constant which must be experimentally determined for each radioactive nuclide. The minus sign is required because the decay rate decreases with time. After integration of (II.2) we obtain:

$$N = N_0 e^{-\lambda t}. \tag{II.3}$$

N_0 is the number of radioactive parent atoms present at the time of formation of a mineral or rock, while t measures the time elapsed since their formation.

It is convenient to describe the rate of decay by means of the "half-life" $(T_{1/2})$, which is defined as the time required for one-half of the parent atoms to decay. Setting $N = N_0/2$ and $t = T_{1/2}$, we find from (II.3) that $e^{\lambda T_{1/2}} = 2$. Taking logarithms to the base e of both sides, we find that

$$T_{1/2} = \frac{\ln 2}{\lambda}. \tag{II.4}$$

Since $\ln 2 = 0.693$, we can easily convert the decay constant of a radioactive nuclide into its corresponding half-life, or vice versa. The decay of radioactive ^{87}Rb and the growth of stable ^{87}Sr is illustrated in Fig. II.1.

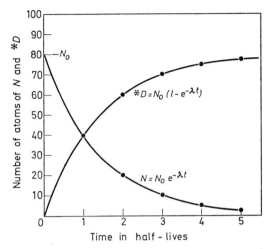

Fig. II.1. Decay of a radioactive parent N and growth of its stable daughter $*D$. The diagram shows the exponential decay of the parent, whose initial abundance is N_0, and the consequent growth of its radiogenic daughter $*D$. Time is measured in half-lives of the parent isotope

3. The Half-life of ⁸⁷Rb

The determination of the decay constant or half-life of ⁸⁷Rb has been difficult because of the low kinetic energy of the beta particles emitted by this isotope. Three different methods have been used, and about 30 different values have been reported. Many of these have been tabulated by LEUTZ, WENNINGER, and ZIEGLER (1962, p. 415), HAMILTON (1965, p. 87), and HEIER and ADAMS (1965). The methods which have been employed can be classified into three categories: (1) direct measurement of the specific beta activity of ⁸⁷Rb; (2) determination of the amount of ⁸⁷Sr produced by decay of ⁸⁷Rb in the laboratory; and (3) comparison of Rb-Sr dates of minerals with concordant U-Pb dates of coexisting minerals, or with K-Ar dates of the same minerals.

The results obtained by these methods since about 1950 suggest half-lifes ranging from $(4.2 \pm 0.4) \times 10^{10}$ years (GEESE-BÄHNISCH and HUSTER, 1954) to $(6.23 \pm 0.3) \times 10^{10}$ years (MACGREGOR and WIEDENBECK, 1954). The most widely-accepted direct determination of the half-life of ⁸⁷Rb is that of FLYNN and GLENDENIN (1959), who reported a value of $(4.70 \pm 0.05) \times 10^{10}$ years, based on the beta-activity of ⁸⁷Rb dissolved in a liquid scintillator. KOVACH (1964) obtained $(4.77 \pm 0.10) \times 10^{10}$ by an identical method. McMULLEN, FRITZE, and TOMLINSON (1966) measured the amount of radiogenic ⁸⁷Sr produced during a period of seven years in a small quantity of purified rubidium perchlorate, and calculated a value of (4.72 ± 0.04)

$\times 10^{10}$ years for the half-life of ^{87}Rb. This value is in excellent agreement
with that reported earlier by FLYNN and GLENDENIN. Nevertheless, most
isotope geologists are using a value $(5.0 \pm 0.2) \times 10^{10}$ years which was
obtained by ALDRICH et al. (1956) from a comparison of ^{87}Sr*/^{87}Rb ratios
of micas with concordant U-Pb dates on coexisting uraninites and monazites
from six pegmatites (^{87}Sr* = radiogenic ^{87}Sr). In this book we shall use a
geologically determined value of the decay constant ($\lambda = 1.39 \times 10^{-11}$ yr^{-1})
that was reported by ALDRICH and his colleagues, because this value seems
to make Rb-Sr dates agree with dates determined by other methods, where
such concordance is to be expected on geologic grounds. However, we
should keep in mind that dates determined on the basis of this decay constant may be systematically too high by about 6 percent, compared to the
best direct measurements of this constant.

4. The Growth of Radiogenic ^{87}Sr in Rocks and Minerals

The number of radiogenic daughter atoms $*D$ which have formed by
decay of parent N in a specimen of a rock or mineral since the date of its
formation t years ago is

$$*D = N_0 - N. \tag{II.5}$$

The total number of daughter atoms in the specimen will be

$$D = D_0 + *D, \tag{II.6}$$

where D_0 is the number of daughter atoms incorporated into the specimen
at the time of its formation and D is the total number of daughters at the
present time. Combining Eq. (II.5) and (II.6) we have

$$D = D_0 + N_0 - N, \tag{II.7}$$

and since $N_0 = Ne^{\lambda t}$ from Eq. (II.3),

$$D = D_0 + N(e^{\lambda t} - 1). \tag{II.8}$$

Eq. (II.8) is applicable to any decay scheme involving the decay of a radioactive parent to a stable daughter. In order to adapt it to the decay of ^{87}Rb
to ^{87}Sr, we substitute the appropriate symbols for D and N:

$$^{87}\text{Sr} = {}^{87}\text{Sr}_0 + {}^{87}\text{Rb}\,(e^{\lambda t} - 1). \tag{II.9}$$

It is useful to divide Eq. (II.9) by the number of ^{86}Sr atoms to form ratios
which are more easily determined in the laboratory than is the number of
atoms of ^{87}Sr or ^{87}Rb in a sample of mineral or rock. This is mathematically
correct because the total number of ^{86}Sr atoms in a sample of rock or mineral
remains constant, while the number of ^{87}Sr atoms increases by decay of
^{87}Rb. Therefore, we may write Eq. (II.9) as

$$\frac{^{87}\text{Sr}}{^{86}\text{Sr}} = \left(\frac{^{87}\text{Sr}}{^{86}\text{Sr}}\right)_0 + \frac{^{87}\text{Rb}}{^{86}\text{Sr}}\,(e^{\lambda t} - 1), \tag{II.10}$$

where

$\dfrac{^{87}Sr}{^{86}Sr}$ = the ratio of these isotopes in the sample at the time of analysis;

$\left(\dfrac{^{87}Sr}{^{86}Sr}\right)_0$ = the same, at the time of formation of the rock or mineral specimen;

$\dfrac{^{87}Rb}{^{86}Sr}$ = the ratio of these isotopes in the sample at the time of analysis; and

t = the time elapsed since last crystallization or isotopic homogenization of the sample.

The $^{87}Sr/^{86}Sr$ ratio of a rock or mineral can be measured using a suitable mass spectrometer. Concentrations of rubidium and strontium can be determined by a variety of techniques, and the $^{87}Rb/^{86}Sr$ ratio can then be calculated. If the initial $^{87}Sr/^{86}Sr$ ratio is known, or if a reasonable value is assumed, Eq. (II.10) can be solved for the variable t.

The resulting date will be the "age" of the rock or mineral, provided the following assumptions are satisfied: (1) the $^{87}Sr/^{86}Sr$ ratio has changed only by decay or ^{87}Rb to ^{87}Sr and not as a result of migration of rubidium or strontium into or out of the rock or mineral; (2) the value of the decay constant is known; and (3) the analyses are accurate and sufficiently precise to make a meaningful age determination possible.

5. The Rubidium-Strontium Isochron Method of Dating

Rubidium-rich minerals such as lepidolite, muscovite, biotite, or potassium feldspar, of Paleozoic age or older, are now dateable by the Rb-Sr method without undue difficulty. Because of their relatively high Rb/Sr ratios, the strontium in these minerals is strongly enriched in radiogenic ^{87}Sr. In such cases the date calculated from Eq. (II.10) is insensitive to the value of the initial $^{87}Sr/^{86}Sr$ ratio, and reasonable estimates of this ratio may be made from a knowledge of the $^{87}Sr/^{86}Sr$ ratios of recent volcanic rocks or of modern marine strontium. Igneous and metamorphic rocks, on the other hand, generally have much lower Rb/Sr ratios than the minerals listed above, and dating by the "whole-rock" method requires a more careful evaluation of initial $^{87}Sr/^{86}Sr$ ratios.

Let us consider the simple case of a suite of comagmatic igneous rocks which formed by fractionation and crystallization of a magma. All specimens of rock belonging to such a comagmatic suite have very nearly the same age. If the strontium in the parent magma had been isotopically homogeneous, the rocks which form from it by magmatic differentiation would also have had the same initial $^{87}Sr/^{86}Sr$ ratio. However, because they have different mineral compositions, different members of a comagmatic suite will in general have different concentrations of rubidium and strontium, and therefore different $^{87}Rb/^{86}Sr$ ratios.

Examination of Eq. (II.10) shows that when t is constant, it reduces to the equation of a family of straight lines in the slope-intercept form

$$y = b + mx .\tag{II.11}$$

The slope of the resulting straight line is

$$m = e^{\lambda t} - 1 ,\tag{II.12}$$

while the intercept is

$$b = \left(\frac{^{87}\text{Sr}}{^{86}\text{Sr}}\right)_0 .\tag{II.13}$$

Consequently, all members of a comagmatic suite of igneous rocks which satisfy the condition of having the same age, the same initial $^{87}\text{Sr}/^{86}\text{Sr}$ ratio, and have remained closed to rubidium and strontium satisfy Eq. (II.10), and therefore form a straight line in coordinates of $^{87}\text{Sr}/^{86}\text{Sr}$ (y) and $^{87}\text{Rb}/^{86}\text{Sr}$ (x). This line is called the *isochron*. The slope of the isochron is directly related to the age of the rocks on the isochron, as shown in Eq. (II.12), while its intercept is equal to the initial $^{87}\text{Sr}/^{86}\text{Sr}$ ratio of the rocks [Eq. (II.13)]. Fig. II.2 is an illustration of this method of dating, which was first proposed by NICOLAYSEN (1961). Further discussion of the isochron is given in the caption to Fig. II.2.

When the $^{87}\text{Sr}/^{86}\text{Sr}$ and $^{87}\text{Rb}/^{86}\text{Sr}$ ratios of rocks or minerals are plotted to scale on an isochron diagram, each point will move as a function of time

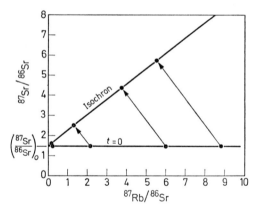

Fig. II.2. An isochron diagram showing the straight-line paths followed by a suite of comagmatic rock or mineral samples. At the time of crystallization ($t = 0$), all samples of a comagmatic suite of igneous rocks are assumed to have the same $^{87}\text{Sr}/^{86}\text{Sr}$ ratio, but may have different $^{87}\text{Rb}/^{86}\text{Sr}$ ratios, depending on their mineral composition. Thereafter, the points representing rocks or minerals move along straight lines having a slope of -1. As long as the rocks or minerals remain closed to rubidium and strontium, they satisfy equation (II.10) and therefore lie on the isochron. The slope of this isochron is related to the age of the rocks by equation (II.12), while their initial $^{87}\text{Sr}/^{86}\text{Sr}$ ratio is given by the intercept of the isochron with the vertical axis

along a straight line having a slope of -1. This can be understood by remembering that each decay reduces the number of ^{87}Rb atoms by one and increases the number of ^{87}Sr atoms by one. However, because of the very long half-life of ^{87}Rb, the resulting change in the ^{87}Sr/^{86}Sr ratio is very small, and it is therefore customary to exaggerate the vertical scale in order to magnify the slope of the isochron. The path taken by points on a conventional isochron diagram is thus much more vertical than shown in Fig. II.2.

In order to date a suite of comagmatic rocks or minerals it is necessary to measure their ^{87}Sr/^{86}Sr and ^{87}Rb/^{86}Sr ratios. These will plot along a straight-line isochron, provided the specimens are in fact comagmatic and provided all other assumptions are satisfied. The goodness of fit of the data points to a single isochron can be used to confirm the assumption that the specimens are comagmatic. This assumption must be based initially on geological information, but can be tested statistically. The age of the specimens forming the isochron can be calculated from its slope using Eq. (II.12):

$$t = 1/\lambda \ln (m + 1) .\qquad\qquad (\text{II.14})$$

Several statistical methods are available for estimating the best slope and zero intercept of isochrons. (YOUDEN, 1951; McINTIRE et al., 1966; YORK, 1966, 1967). This method of dating thus provides not only the age of a comagmatic suite of rocks but also their initial ^{87}Sr/^{86}Sr ratio.

Before we leave the subject of age determinations, a comment is in order regarding the use of the terms "date" and "age." We suggest that a "date" is an instant of time in the geologic past that can be specified in terms of the amount of time that has passed since then. An "age," on the other hand, refers to the amount of time that has passed since the occurrence of a specific geologic event, such as the formation of a rock or mineral. Thus the term "age" is restricted to those "dates" when significant geological events occurred. The solutions of Eqs. (II.10) or (II.14) are considered to be "dates" and are given the status of "age" only after subsequent evaluation of all available and pertinent evidence indicates that the "date" does, in fact, mark the occurrence of a geological event. The distinction between "dates" and "ages" is complicated by the fact that several different dateable events may be recorded by the isotopic systems within the minerals of a rock.

6. A Useful Approximation

It is sometimes sufficient for the purpose of making rapid and approximate calculations to modify Eq. (II.10) in the following way. Let

$$e^{\lambda t} = 1 + \lambda t + \frac{(\lambda t)^2}{2!} + \frac{(\lambda t)^3}{3!} + \cdots .$$

If λt is small, as in the case of the decay of ^{87}Rb, then

$$1 + \lambda t \gg \frac{(\lambda t)^2}{2!} + \frac{(\lambda t)^3}{3!} + \cdots$$

and

$$e^{\lambda t} \cong 1 + \lambda t.$$

Consequently, to a good approximation:

$$\frac{^{87}Sr}{^{86}Sr} \cong \left(\frac{^{87}Sr}{^{86}Sr}\right)_0 + \frac{^{87}Rb}{^{86}Sr} \lambda t \qquad (II.15)$$

and

$$\frac{^{87}Sr}{^{86}Sr} \cong \left(\frac{^{87}Sr}{^{86}Sr}\right)_0 + \frac{Rb}{Sr} k \lambda t, \qquad (II.16)$$

where

$$k = \frac{(Ab\ ^{87}Rb)\ (at.\ wgt.\ Sr)}{(Ab\ ^{86}Sr)\ (at.\ wgt.\ Rb)}$$

and $(Ab\ ^{87}Rb)$ = isotopic abundance of ^{87}Rb in atom percent, etc. The numerical value of k depends on the atomic weight of strontium and on the isotopic abundance of ^{86}Sr $(Ab\ ^{86}Sr)$ both of which depend on the ^{87}Sr-enrichment, and thus on the $^{87}Sr/^{86}Sr$ ratio, of a particular sample. For a sample having a $^{87}Sr/^{86}Sr$ ratio of 0.7093 (modern seawater) we find that $k = 2.8936$, assuming that the abundance of ^{87}Rb is 27.8346 percent and

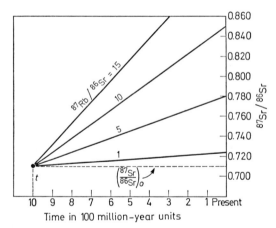

Time in 100 million-year units

Fig. II.3. The increase of $^{87}Sr/^{86}Sr$ ratios as a function of time in a suite of comagmatic rocks or minerals. As time passes, the $^{87}Sr/^{86}Sr$ ratio of each rock or mineral of a comagmatic suite increases linearly, as described by equation (II.15). The slopes of the development lines are equal to $(^{87}Rb/^{86}Sr)\ \lambda$, causing the lines to diverge in a fan-shaped array. After $^{87}Sr/^{86}Sr$ and $^{87}Rb/^{86}Sr$ ratios of a suite of comagmatic rocks or mineral have been measured, development lines can be plotted back into the past and will converge to a point, provided the rocks or minerals have remained closed to rubidium and strontium. The coordinates of the point of convergence are the age and initial $^{87}Sr/^{86}Sr$ ratio of the suite

that the atomic weight of rubidium is 85.46776 atomic mass units. In addition, we assume that $^{86}Sr/^{88}Sr = 0.1194$ and $^{84}Sr/^{88}Sr = 0.00675$. The approximate Eq. (II.15) overestimates the value of t compared to the exact form of the Eq. (II.10) by about 2.7 percent for dates of 4.5 billion years. In summary, Eq. (II.15) and (II.16) are useful approximations, but should be used only when accurate results are not required.

Eq. (II.15) represents a family of straight lines in coordinates of $^{87}Sr/^{86}Sr$ and t. A suite of comagmatic igneous rocks having a common initial $^{87}Sr/^{86}Sr$ ratio but differing $^{87}Rb/^{86}Sr$ ratios will be represented on a plot of $^{87}Sr/^{86}Sr$ versus t by a fan-shaped array of straight lines whose slopes are given by $(^{87}Rb/^{86}Sr)\,\lambda$. Fig. II.3 is such a family of straight lines which are known as "development lines," because each line shows how the $^{87}Sr/^{86}Sr$ ratio has developed in a particular sample as a function of time.

7. Analytical Methods

Concentrations of rubidium and strontium in rocks and minerals can be measured in several different ways. STRASSMAN and WALLING (1938) and STEVENS and SCHALLER (1942) used gravimetric methods. AHRENS (1949) proposed an optical spectrographic method. However, these techniques were neither sufficiently accurate nor sensitive and they have been replaced by the isotope dilution method introduced by TOMLINSON and DAS GUPTA (1953) and ALDRICH et al. (1953a, b). More recently FAIRBAIRN et al. (1967), PETERMAN et al. (1968), KROGH and HURLEY (1968), and POWELL et al. (1969) have used X-ray fluorescence spectrometry, discussed in detail by NORRISH and CHAPPELL (1967). Isotope dilution is the most sensitive and precise of the available techniques, primarily because it is entirely free of interference from other elements. However, many investigators are now using X-ray fluorescence to determine concentrations of rubidium and strontium for dating of rocks by the Rb-Sr isochron method.

To allow for determination of the concentration of an element by isotope dilution an element must, in general, have two stable isotopes. In addition, it must be possible to obtain a sample of the element (the *spike*) in which one of the stable isotopes has been greatly increased in abundance. In general, a known weight (or volume) of an aqueous solution containing a known amount of spike of known isotopic composition is added to a known weight of the powdered sample. The rock or mineral powder is digested in a mixture of hydrofluoric acid and sulfuric or perchloric acid. The residue is then dissolved in dilute hydrochloric acid. During the digestion and solution of the samples the spike and the normal element in the sample become intimately mixed. The isotopic composition of the element in the mixture will then be intermediate between that of the spike and the normal element in the sample. For isotope analyses of rubidium the spike is enriched in ^{87}Rb. In the case of strontium, one may use spikes enriched in ^{84}Sr or ^{86}Sr.

The principle of isotope dilution described above is further illustrated in Fig. II.4a, b, c, which show mass spectra of normal rubidium (a), spike rubidium enriched in ⁸⁷Rb (b), and a mixture of normal and spike rubidium (c), prepared in order to determine the rubidium content of a rock.

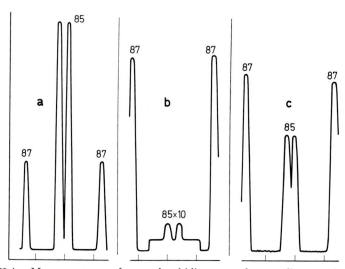

Fig. II.4a. Mass spectrum of natural rubidium traced on a linear strip chart recorder at a constant scan rate. The height of the peaks is proportional to the abundance of the isotope at each mass. In natural rubidium the abundance of ⁸⁵Rb is 72.1654, while that of ⁸⁷Rb is 27.8346. (Data from CATANZARO et al., 1969)

Fig. II.4b. Mass spectrum of spike rubidium which has been enriched in ⁸⁷Rb by a factor of about 3.6 compared to natural rubidium. The abundance of ⁸⁷Rb is 99.40 percent, while that of ⁸⁵Rb is 0.60 percent. Note that the ⁸⁵Rb peak was magnified 10 times compared to ⁸⁷Rb

Fig. II.4c. Mass spectrum of a mixture of natural and spike rubidium. This is an actual record of a rubidium determination in a Cambrian rhyolite from the Long Hills in the Transantarctic Mountains. Note that the abundances of the isotopes in this mixture are intermediate between those of natural and spike rubidium. The concentration of rubidium in this rock can be calculated from the observed ⁸⁷Rb/⁸⁵Rb ratio of the mixture and from knowledge of the isotopic compositions of natural and spike rubidium, the weight of spike rubidium added to the sample, and the weight of the sample

After the sample has been dissolved and the isotopes have equilibrated, rubidium and strontium are separated from each other and from the other cations by passing the solution through a cation exchange column and eluting with dilute hydrochloric acid. Quantitative recovery of the element to be determined from the ion-exchange column is not required. After the spike and the normal element in the sample have completely mixed, any fraction of the mixture can be analyzed on a suitable mass spectrometer.

From the observed isotopic composition of the mixture the concentration of the element in the sample can then be calculated. The necessary calculations have been described in detail by HAMILTON (1965, pp. 10—15) and WEBSTER (1960), and need not be repeated here. LONG (1966), BOELRIJK (1968), and KROGH and HURLEY (1968) have published discussions of the theoretical aspects of the determination of normal and radiogenic strontium by isotope dilution using ^{84}Sr-enriched spike.

Age determinations by the Rb-Sr method also require a measurement of the ^{87}Sr/^{86}Sr ratio of the rock or mineral on a mass spectrometer equipped with a suitable ion source. The design of mass spectrometers used in geochronometry has been reviewed by HAMILTON (1965, pp. 15—31), MAYNE (1960, pp. 177—195), FRIEDMAN (1954, pp. 64—70), and others. More general treatments of the principles of mass spectrometers are available in publications by INGHRAM (1948), DUCKWORTH (1960), and McDOWELL (1963). Fig. II.5 is a photograph of a mass spectrometer designed for mass analysis of solid samples. Instruments such as this one are widely used for dating by the Rb-Sr method.

The rock or mineral powder is dissolved as described earlier. The strontium is separated from the other elements by cation exchange chromatography and is placed as a salt on a ribbon of tantalum or rhenium in the

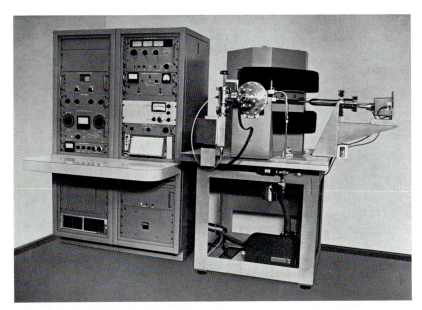

Fig. II.5. Solid-source mass spectrometers such as this Model 12-90-S of Nuclide Corporation of State College, Pennsylvania, are widely used for mass analysis of strontium in geological materials

source of the mass spectrometer. The ribbon containing the sample, or an adjacent "ionizing" ribbon, is heated in the evacuated mass spectrometer to produce Sr^{1+} ions by thermionic emission. The ions are accelerated by an electric field through a series of slit-plates in the source to form a collimated beam of positively-charged ions.

The ion beam then enters the magnetic field of the electromagnetic analyzer, causing the ions to be deflected. The heavier isotopes are deflected less than the lighter ones, and in this way the ion beam is separated into its components to produce a mass spectrum. The separated ion beams emerging from the magnetic field continue to the collector, where they are allowed to strike a shielded cup. One or several slit-plates are arranged in front of the collector in such a way that only one isotopic ion beam can enter the shielded cup at any time. The ions striking the cup are neutralized by electrons which flow to the cup from ground through a resistor (10^9 to 10^{12} ohms). The potential difference generated across the terminals of this resistor by the electron current is magnified and measured by a vibrating reed electrometer (V.R.E.). The output from the V.R.E. is recorded in several alternate ways, such as potentiometric strip-chart recorders or integrating digital voltmeters. The magnetic field strength is changed either continually or in steps by variation of the current in the coils of the electromagnet in such a way that the isotopic beams are focused into the collector

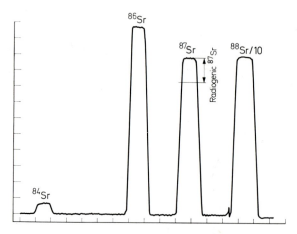

Fig. II.6. This mass spectrum of unspiked strontium was obtained in the Laboratory for Isotope Geology and Geochemistry at The Ohio State University on a Nuclide Corp. Model 6-60-S mass spectrometer, using a constant scan rate. The strontium was extracted from a Cambrian rhyolite collected in the Long Hills of the Transantarctic Mountains. The $^{87}Sr/^{86}Sr$ ratio of this strontium is 0.828, which means that about 15.6 percent of the ^{87}Sr atoms are radiogenic and have formed by decay of ^{87}Rb in this rock specimen

cup in succession. If the ion beams are swept continuously past the collector cup, a mass spectrum of the element being analyzed is traced out on a strip-chart recorder as a series of peaks. Each peak corresponds to a particular isotope, and its height is proportional to the abundance of that isotope. Fig. II.6 is such a mass spectrum of strontium extracted from a Cambrian rhyolite in the Long Hills of the Transantarctic Mountains.

The foregoing is a very general description of the determination of the $^{87}Sr/^{86}Sr$ ratios in rocks and minerals. More sophisticated instrumentation is in use in several laboratories, but the technical details go beyond the scope of this book.

An important constraint in the interpretation of $^{87}Sr/^{86}Sr$ ratios in geological materials is the possibility that strontium isotopes may be fractionated by geologic processes, such as partial melting of silicate rocks, crystallization of magma, diffusion, or other processes. Actually, no instance of fractionation of strontium isotopes in nature has yet been reported. One possible reason for this is that the isotopes of strontium are fractionated in the mass spectrometer to an extent that obscures any original or natural fractionation. In order to minimize the effects of fractionation on the measured $^{87}Sr/^{86}Sr$ ratios, a correction is now routinely made to all analyses. This correction was first used by HERZOG et al. (1958, p. 721) and FAURE and HURLEY (1963, p. 34). It is based on the assumption that the $^{86}Sr/^{88}Sr$ ratio is equal to 0.1194 and that the fractionation of two isotopes is proportional to the difference in their masses. Consequently, the measured $^{87}Sr/^{86}Sr$ ratios are adjusted only one-half as much as required to change the $^{86}Sr/^{88}Sr$ ratio from its measured value to the assumed value of 0.1194. The correction factor f by which the measured $^{87}Sr/^{86}Sr$ ratios are multiplied is:

$$f = \frac{M}{\dfrac{M + 0.1194}{2}} = \frac{2M}{M + 0.1194}, \tag{II.17}$$

where M is the measured $^{86}Sr/^{88}Sr$ ratio. This correction reduces the effects of both natural and instrumental fractionation and generally improves the reproducibility of $^{87}Sr/^{86}Sr$ ratios.

The analytical uncertainty of $^{87}Sr/^{86}Sr$ ratios, corrected for fractionation, is ± 0.1 percent, or better. The $^{87}Rb/^{86}Sr$ ratios of rocks and minerals generally have analytical uncertainties of about ± 2 percent, but they may be larger or smaller than this depending on the absolute concentrations of rubidium and strontium.

8. Summary

The decay of naturally-occurring ^{87}Rb to stable ^{87}Sr can be used to measure the ages of a variety of rocks and minerals which contain rubidium.

The most useful method of dating is based on isochrons formed in co-ordinates of $^{87}Sr/^{86}Sr$ and $^{87}Rb/^{86}Sr$ by cogenetic rocks and minerals having the same age and initial $^{87}Sr/^{86}Sr$ ratio. The "dates" indicated by the solutions of the radioactivity equation are regarded as "ages" only when they can be associated with a specific geological event. The analytical techniques and instrumentation required for dating by the Rb-Sr method are capable of great precision and accuracy and make possible the dating of a great variety of rocks and minerals.

III. Uses of Strontium Isotopes in Petrogenesis

1. Introduction

The isotopic composition of strontium is not only a useful indicator of the ages of rocks and minerals, but it also contains information about the origin of igneous rocks and about the geologic processes that have affected their chemical compositions. We showed in the previous chapter that the Rb-Sr isochron method of dating leads not only to the determination of the age of a suite of comagmatic igneous rocks, but also indicates the "initial ratio," which is the $^{87}Sr/^{86}Sr$ ratio of the magma from which the rocks crystallized. Its value depends on the previous history of the strontium and reflects particularly the Rb/Sr ratios of systems in which the strontium had previously resided. In general there is no way of knowing, in a particular case, in how many systems the strontium previously resided, what the Rb/Sr ratios were in those systems, or how long the strontium resided in each system. We can deal rigorously only with single-stage histories by assuming that the strontium was part of only one previous system for a specified period of time before being incorporated into a particular magma. Although these assumptions undoubtedly oversimplify the real situation in many instances, the approach has shed light on the origin of igneous rocks.

In this chapter we shall outline the assumptions that will permit us in the following chapters to use the initial $^{87}Sr/^{86}Sr$ ratios of plutonic and volcanic igneous rocks to suggest where a particular magma originated in the interior of the Earth and whether it was subsequently modified chemically by interactions with other rocks.

2. Evolution of Strontium in the Continental Crust and the Upper Mantle

One of the most important questions in petrology is whether magma is generated exclusively in the upper mantle of the Earth or whether certain kinds of magma originate (either exclusively or in part) within the continental crust. If magma-generation takes place in both places, can we determine in which place a particular magma was formed? We shall show now that ^{87}Sr can be used as a geologic tracer to identify the sources of magma.

Before proceeding, we must clarify what we mean by the terms "upper mantle" and "continental crust." The upper mantle extends from the Mohorovičić discontinuity to a depth of about 600 kilometers below the surface of the Earth. Melting in this region and the upward migration of the resulting magma have contributed to the formation of the crust of the Earth throughout geologic time. The continental crust includes the rocks which lie above the Mohorovičić discontinuity in the continents. It is distinguished from the "oceanic crust" by the average chemical composition and age of its rocks and by its greater thickness.

It is well-established that the rocks of the upper mantle of the Earth are composed primarily of silicates of iron and magnesium. Compared to the upper mantle, the rocks of the continental crust are enriched in silica, alumina, and the alkali metals. In keeping with the geochemical properties of rubidium and strontium outlined in Chapter I, the rocks of the continental crust are enriched in rubidium and therefore have significantly higher Rb/Sr ratios than the rocks of the upper mantle. Consequently, strontium in the rocks of the continental crust with time has become enriched in radiogenic ^{87}Sr compared to strontium in the upper mantle. It is therefore reasonable to postulate that the $^{87}Sr/^{86}Sr$ ratios of granitic rocks in the continental crust are significantly and measurably greater than those of the rocks in the upper mantle.

The isotopic composition of strontium in the upper mantle can be measured directly by analysis of basaltic rocks extruded in the ocean basins or in the continents, providing that each of the following conditions holds: (1) the magma was generated in the upper mantle; (2) strontium in the magma was not contaminated with foreign strontium derived from another source; and (3) the strontium in the magma was isotopically identical to that of the solid mantle from which the magma was generated. The first condition can be shown to hold for oceanic volcanic rocks, but the second must be evaluated in each case, and no generalizations are possible. The third condition implies the assumption that the isotopes of strontium are not fractionated by melting of rocks in the upper mantle, by subsequent chemical reactions, or by crystallization of minerals from the magma. No evidence has ever been presented to suggest that strontium isotopes are fractionated in nature, and it is therefore widely believed that fractionation effects are negligible. In any case, the correction that is made to all measured $^{87}Sr/^{86}Sr$ ratios (Chapter II) corrects for the effects of fractionation occurring in the mass spectrometer and also for any that might have occurred in nature.

The $^{87}Sr/^{86}Sr$ ratios of oceanic-island basalts, to be discussed in more detail in Chapter IV, range from about 0.702 to 0.706, and average 0.7037. Analyses of strontium in stony meteorites (basaltic achondrites) can be used to suggest that the $^{87}Sr/^{86}Sr$ ratio of the Earth about 4.6 billion years ago had a value close to 0.699 (see Chapter X). With this information we can calculate the average present-day Rb/Sr ratio of the upper mantle sources of the island basalts, assuming that the strontium of Recent oceanic basalts resided in the upper mantle for 4.6 billion years and that its primordial $^{87}Sr/^{86}Sr$ ratio was 0.699. Using Eq. (II.16), derived in Chapter II, and substituting $^{87}Sr/^{86}Sr = 0.7037$, $(^{87}Sr/^{86}Sr)_0 = 0.699$, and $t = 4.6 \times 10^8$ years, we find that the average Rb/Sr ratio of the source regions of island basalt magma in the upper mantle is 0.025.

The average $^{87}Sr/^{86}Sr$ and Rb/Sr ratio of the granitic rocks of the continental crust is more difficult to determine. As geologists, we are well aware of the great diversity of chemical and mineralogic compositions and of the great range of the ages of rocks on the continents. Several investigators have attempted to measure the average $^{87}Sr/^{86}Sr$ ratios of the continental crust by analyzing strontium in water or in mollusk shells of lakes and rivers draining large areas of the continents. The available data are compiled in Table III.1. They demonstrate that $^{87}Sr/^{86}Sr$ ratios of strontium released by weathering of old igneous and metamorphic rocks are clearly higher than those of Recent oceanic basalts. It is also apparent, however, that the $^{87}Sr/^{86}Sr$ ratio of the continental crust is quite variable and depends on the Rb/Sr ratios and the ages of the rocks exposed in the drainage basins that have been sampled.

For the purpose of this discussion it is sufficient to derive an approximate average $^{87}Sr/^{86}Sr$ ratio for the sialic rocks of the continental crust from a simple calculation similar to that made for the upper mantle before. We assume that on the average the sialic rocks of the continental crust are 2.5 billion years old and that the strontium they contain was derived from the upper mantle at the time of formation of these rocks. According to the model for strontium evolution presented above, the average $^{87}Sr/^{86}Sr$ of the upper mantle at that time was 0.701. The compilations of trace-element concentrations in rocks by TAYLOR (1965), (see Chapter I) lead to an estimated value of 0.24 for the Rb/Sr ratio of the crust. We prefer a lower value of 0.18 because it is more representative of rocks exposed on the Precambrian Shield of North America and elsewhere. Using Eq. (II.16) and setting $(^{87}Sr/^{86}Sr)_0 = 0.701$, Rb/Sr = 0.18, and $t = 2.5$ billion years, we obtain an

Table III.1. The isotopic composition of strontium in lakes and rivers draining areas composed mainly of silicate rocks

Source of Strontium	$^{87}Sr/^{86}Sr$	Reference
Lakes and rivers, Canadian Precambrian Shield	0.712—0.726	FAURE, HURLEY, and FAIRBAIRN (1963)
Lake Superior, North America	0.718	HART and TILTON (1966)
Lake Vanda, Wright Valley, Southern Victoria Land, Antarctica	0.7149	JONES and FAURE (1967)
Lake Bonney, Taylor Valley, Southern Victoria Land, Antarctica	0.7136	JONES and FAURE (1968)
Lake George, near Killarney, Ontario, Canada	0.7184	JONES and FAURE (unpublished)
Great Salt Lake, Utah, USA	0.7174	JONES and FAURE (unpublished)

estimate of 0.719 for the average ^{87}Sr/^{86}Sr ratio of strontium in the continental crust. This value is consistent with observed isotopic compositions of strontium in surface water on the continents shown in Table III.1.

The isotopic evolution of strontium in the upper mantle and the continental crust is summarized as a development diagram in Fig. III.1. It will serve as the basis for the interpretation of initial ^{87}Sr/^{86}Sr ratios of igneous rocks outlined in the next section.

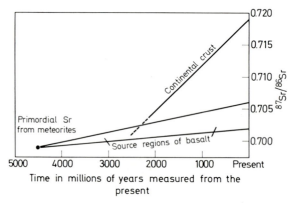

Fig. III.1. The evolution of strontium in the source regions of oceanic basalt (upper mantle) and the continental crust. Strontium in the continental crust has developed a higher ^{87}Sr/^{86}Sr ratio than the source regions of oceanic basalt because its Rb/Sr ratio is greater than that of the upper mantle. FAURE and HURLEY (1963) suggested that igneous rocks which crystallized from magma that was derived from the upper mantle can be recognized because they will have initial ^{87}Sr/^{86}Sr ratios falling within the "basalt field"

3. Applications to Petrology

We are now ready to outline the use of the initial ^{87}Sr/^{86}Sr ratios of igneous rocks to determine their petrogenesis. When magma is generated in the upper mantle and solidifies in the crust or on the surface of the Earth without contamination with foreign strontium, the resulting rocks should have an initial ^{87}Sr/^{86}Sr ratio that lies within the "basalt field" outlined in Fig. III.1. On the other hand, when magma is generated by partial melting of old granitic rocks in the continental crust, the rocks formed from it are expected to have an initial ^{87}Sr/^{86}Sr ratio significantly higher than those of the source regions of basalt. This principle was proposed by FAURE and HURLEY (1963). It has direct bearing on the long-standing controversy regarding the origin of granite, on the petrogenesis of andesite, and on related problems involving the sources of magma from which different kinds of igneous rocks crystallize. These will be examined in more detail in succeeding chapters.

An equally important problem in petrology concerns the process whereby a single magma may crystallize to produce a suite of rocks having a range of chemical compositions. Traditionally petrologists have explained the differentiation of magma by appealing to fractional crystallization, as developed with great clarity by N. L. BOWEN in his book: "The Evolution of the Igneous Rocks."

When magma undergoes fractional crystallization in a closed system, that is, without incorporation of foreign strontium from the country rock, all of the resulting rocks should have the same initial $^{87}Sr/^{86}Sr$ ratio, regardless of their chemical compositions. On the other hand, if foreign strontium entered the magma as crystallization progressed, the rocks formed at different stages during this process may have different initial $^{87}Sr/^{86}Sr$ ratios. Several examples of this phenomenon have now come to light which suggest that the chemical differentiation of some suites of volcanic rocks is due not only to fractional crystallization of the parent magma, but also to extensive interactions of the magma with country rock.

4. Summary

Because the Rb/Sr ratio of the continental crust is about 10 times greater than that of the upper mantle, strontium in the continental crust has become enriched in ^{87}Sr. Therefore, rocks formed by melting, metasomatism, or assimilation of typical crustal materials will be labeled by having higher initial $^{87}Sr/^{86}Sr$ ratios than the ratios of uncontaminated rocks derived from the mantle.

IV. Volcanic Rocks

1. Introduction

The study of radiogenic isotopes in oceanic volcanic rocks provides one of the best ways of obtaining information about the upper mantle. Since the effects of fractionation of the strontium isotopes can be removed (see Chapter II), the initial $^{87}Sr/^{86}Sr$ ratio of an uncontaminated oceanic basalt will be the same as that of its source region in the upper mantle. The present $^{87}Sr/^{86}Sr$ ratio of any part of the upper mantle depends on its initial ratio, on its average Rb/Sr ratio, and on its age [see Eq. (II.15)]. Therefore, the observed $^{87}Sr/^{86}Sr$ ratios of modern oceanic basalts provide information not only about the isotopic composition of strontium in the upper mantle, but also about its possible Rb/Sr ratio. GAST (1960) and HURLEY (1968a, b), among others, have shown how Rb/Sr ratios can be used as a clue to the chemical composition and evolution of the mantle.

GAST (1967) made a comprehensive review of the isotopic geochemistry of lead, oxygen, strontium, and sulfur in volcanic rocks. His survey of the literature of strontium isotope geology was completed in early 1965 (see GAST, 1967, p. 338), and he listed the $^{87}Sr/^{86}Sr$ ratios of approximately 160 young volcanic rocks (Table II, p. 340). At the time this book is being written, the number of reported analyses has increased more than five-fold, reflecting the greatly increased activity in strontium isotope geology since GAST's survey.

2. The Strontium Isotopic Data for Volcanic Rocks

a) General Comment

In our discussion we shall subdivide modern volcanic rocks into a few principal groups: oceanic-island basalts, sea-floor basalts, circum-oceanic andesitic rocks, Antarctic and Tasmanian Jurassic dolerites, and continental basaltic and felsic volcanic rocks. Fig. IV.1 is a graph of initial $^{87}Sr/^{86}Sr$ ratio versus strontium content for these groups, and for dunites and peridotites whose discussion we defer to Chapter VII. This graph is useful because it illustrates at a glance the range of the initial $^{87}Sr/^{86}Sr$ ratio and the strontium content shown by the groups and because it reveals that overall there is an approximate inverse correlation between the initial $^{87}Sr/^{86}Sr$ ratio and the strontium content. Put another way, the graph illustrates that, relatively speaking, volcanic rocks with high strontium contents have low

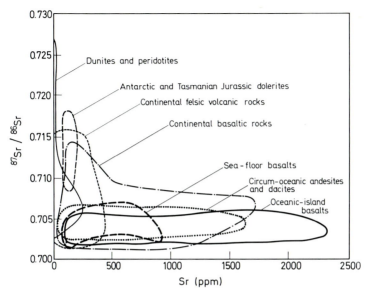

Fig. IV.1. A composite graph showing the range of initial $^{87}Sr/^{86}Sr$ ratio and strontium content for groups of modern volcanic rocks and for dunites and peridotites. The graph is based on a survey of the literature of strontium isotope geology completed in August 1971. Individual data points were plotted for each group, and encircled by a line that included 90 to 100 percent of them. The individual graphs were then superimposed to produce the composite. SINHA and DAVIS (1971) used a similar diagram. *Oceanic Island Basalts* ($N = 133$) include basaltic rocks from oceanic islands. A disproportionately large number are from the Hawaiian Islands. *Sea-Floor Basalts* ($N = 42$) include basalts dredged from the sea floor. *Circum-Oceanic Andesites and Dacites* ($N = 67$) include only those occurring in circum-oceanic volcanic arcs. *Continental Felsic Rocks* ($N = 102$) include intermediate and silicic varieties; *Continental Basaltic Rocks* ($N = 212$) include those less silicic than andesite. *Antarctic and Tasmanian Jurassic Dolerites* ($N = 45$) include the Jurassic dolerites of Tasmania and those of the Ferrar group of Antarctica. *Dunites and Peridotites* ($N = 76$) will be discussed in Chapter VII

$^{87}Sr/^{86}Sr$ ratios; and volcanic rocks, as well as dunites and peridotites, with high $^{87}Sr/^{86}Sr$ ratios have low strontium contents. In general, this relationship might be explained in two ways: (1) Rocks with low strontium contents may come from source regions that also had low strontium contents, and therefore relatively high Rb/Sr ratios. These source materials, by virtue of their higher Rb/Sr ratios, have generated larger than average amounts of ^{87}Sr, and therefore melts produced from them have relatively high $^{87}Sr/^{86}Sr$ ratios. (2) The $^{87}Sr/^{86}Sr$ ratios of melts with low strontium contents can be affected more by a given amount of contamination with foreign strontium than those of melts that are richer in strontium. The high

^{87}Sr/^{86}Sr ratios of some rocks low in strontium content, therefore, may have been caused by contamination of their parent melts with crustal strontium. In fact, both mechanism 1 and mechanism 2 could be important, even for an individual volcanic rock.

b) Basaltic Rocks from Oceanic Islands

Oceanic basaltic rocks in general cannot have been significantly contaminated with older sialic material, which is present only in a few isolated places in the ocean basins. Therefore, the initial ^{87}Sr/^{86}Sr ratios of oceanic basalts provide the most reliable indicators of the isotopic composition of strontium in the upper mantle. In addition, as explained in Chapter III, their ratios are used as a reference level to which those of other rocks, suspected of having an origin wholly or in part in the sialic crust, can be compared.

A histogram of the data available for 164 basaltic rocks from oceanic islands appears as part of Fig. IV.2. The island basalts show a rather narrow normal distribution, with a mean initial ^{87}Sr/^{86}Sr ratio of 0.7037 ± 0.0001 ($\bar{\sigma}$). The data shown have not been corrected for interlaboratory discrepancies, as explained in the Preface. HEDGE et al. (1970) reported that the average ^{87}Sr/^{86}Sr ratio of 90 oceanic island basalts, each corrected to 0.7080 for the Eimer and Amend interlaboratory strontium standard, is 0.7035.

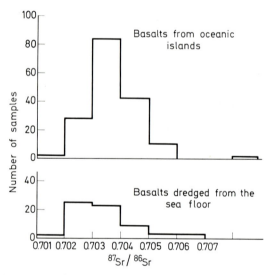

Fig. IV.2. A histogram of the ^{87}Sr/^{86}Sr ratios of basalts from oceanic islands and from the sea floor. More are shown than in Fig. IV.1 because a number of ^{87}Sr/^{86}Sr ratios for basaltic rocks have been reported without accompanying strontium contents. Although the two groups overlap, there is a clear tendency for the sea floor basalts to have lower ratios

Nearly one-half of the island basalts represented in Fig. IV.2 are from the Hawaiian Islands, but in spite of this, the very small range observed indicates that 0.7037 is probably a representative average for the kind of oceanic basaltic rocks that have been analyzed to date. It is obvious that the selection of oceanic rocks in the past has been almost completely dominated by those exposed on islands. We realize now that such volcanic rocks may be less representative of the upper mantle than the volumetrically-more-important abyssal or sea-floor basalts that are being dredged from the ocean floor. In fact, the very existence of a large volcanic pile may indicate that the mantle beneath it is abnormally rich in heat-producing elements — and in their daughter products, including radiogenic lead and strontium isotopes (see ARMSTRONG, 1968). The data available for abyssal or sea-floor basalts will be discussed below, and we shall see that their ratios are often lower than the mean for island basalts. For these reasons the average present-day $^{87}Sr/^{86}Sr$ ratio of the suboceanic upper mantle is certainly no greater than 0.7037, and probably falls in the range of 0.702 to 0.703.

In the last chapter we showed how we can use Eq. (II.16) to calculate the Rb/Sr ratio of the source regions of oceanic island basalts. By setting $(^{87}Sr/^{86}Sr) = 0.7037$, the mean for island basalts; $(^{87}Sr/^{86}Sr)_0 = 0.699$, the meteoritic value (see Chapter X); and $t = 4600$ million years, the "age of the Earth"; we obtained (Rb/Sr) source $= 0.025$. As GAST (1967) pointed out, however, this is a *time-averaged* Rb/Sr ratio; it assumes that the Rb/Sr ratio of the source has remained the same throughout its history — a "single-stage" model. In fact, the Rb/Sr ratio of the source could have been 0.045 for the first 2.3 billion years of its existence, and 0.005 for the last 2.3 billion years, or vice-versa. Obviously there are an infinite number of ways the *time-averaged* Rb/Sr ratio of 0.025 could have been produced. Nevertheless, the figure of 0.025 is fairly typical of the Rb/Sr ratios STUEBER and MURTHY (1966) measured in ultramafic rocks, consistent with the view that the upper mantle is ultramatic in composition.

The *observed* Rb/Sr ratios of oceanic-island basalts are frequently higher than those calculated as above for their source materials. This is to be expected, since during partial melting rubidium is partitioned in the liquid phase more strongly than strontium. But as GAST (1967) noted, there is little evidence that the observed Rb/Sr ratio of an *individual* specimen bears any close or systematic relation to the ratio calculated for its source material from its $^{87}Sr/^{86}Sr$ ratio.

c) Sea-Floor Basalts

One might have thought that all of the important igneous rock-types would have been discovered long ago. And yet, until recently, one major type remained hidden from view on the ocean floor, accessible only by dredging. The distinguishing features of these sea-floor basalts — their tholeiitic composition, their low contents of potassium and other large cations, their high K/Rb ratios, etc. — were established by ENGEL et al. (1965), and the first isotopic measurements were reported by TATSUMOTO et al. (1965). HEDGE and PETERMAN (1970) and PETERMAN and HEDGE (1971) have reviewed the strontium isotope results, and our discussion follows theirs.

It is clear from Fig. IV.2 that although the $^{87}Sr/^{86}Sr$ ratios of the sea-floor basalts cover the same range as those of the island basalts, the distribution is skewed toward slightly lower ratios in the sea-floor basalts. Fig. IV.3, taken from PETERMAN and HEDGE (1971), shows that in a general way the $^{87}Sr/^{86}Sr$ ratios of oceanic basalts, including sea-floor types, correlate positively with their relative potassium contents (defined by $K_2O/K_2O + Na_2O$). This trend probably comes about because a region of the upper mantle that is relatively rich in potassium will also be relatively rich in rubidium, and therefore in time will develop, and produce rocks with higher $^{87}Sr/^{86}Sr$ ratios.

GAST (1968) suggested that the sea-floor basalts, with their relatively unradiogenic strontium, were derived from depleted zones of the mantle that had already produced alkali basalts in a previous episode of melting. PETERMAN and HEDGE explained the trend shown in Fig. IV.3 as follows: (1) the upper mantle has generally lost rubidium to the crust (see HURLEY,

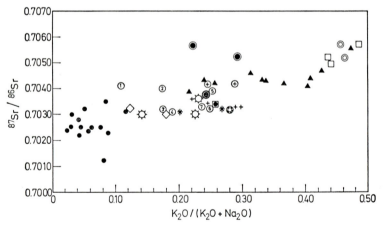

Fig. IV.3. Variations of $^{87}Sr/^{86}Sr$ and $K_2O/(K_2O + Na_2O)$ in oceanic basaltic rocks. The positive correlation is probably caused by the fact that source regions rich in potassium will also be rich in rubidium, and, in time, rich in ^{87}Sr. Key: Ocean ridge tholeiites ● ; Hawaiian Islands ○ : 1 = Koolau Series, 2 = Tholeiites on Hawaii, 3 = Waianae Series, 4 = Honolulu Series, 5 = Kula Series, 6 = Alkaline Series on Maui, 7 = Haulalai Series; Easter ✿; Guadalupe + ; Galapagos ◙ ; Eniwetok ◕ ; Tahiti ▲ ; Samoa ◉ ; Gough □ ; St. Helena ✳ ; Tristan da Cunha ◎ ; Iceland ◇ ; Réunion ⊕ ; Ascension ◑ . (After PETERMAN and HEDGE, 1971)

1968a, b; GAST, 1968), (2) the most rubidium-rich regions of the mantle have suffered the least loss and are in that sense the most primitive, and (3) it follows that the more alkalic oceanic lavas, which are richest in radiogenic strontium, come from the most rubidium-rich zones, and thus sample the most primitive (least-depleted) parts of the upper mantle.

Assuming they are not due to contamination, the differences between the $^{87}Sr/^{86}Sr$ ratios of the sea-floor basalts and those of the island basalts are large enough, in view of the low Rb/Sr ratios of the upper mantle, to require hundreds of millions or billions of years for their generation. Lead isotope variations observed in oceanic lavas likewise require very long times for their development (for example, TATSUMOTO, 1966; OVERSBY and GAST, 1970). These variations suggest that the mantle is heterogeneous, in spite of convection or any other process, even over periods of more than a billion years.

d) Andesites of Volcanic Arcs

Andesites, the characteristic rocks of the arc-trench subduction zones, have received increasing attention with the advent of the theory of plate tectonics. DICKINSON (1970) has given a very comprehensive review of the relationship between andesites and arc-trench tectonics and of the strontium isotope data for andesites. He grouped the theories of origin of andesites into three main categories: (1) primary mantle melts or derivative melts formed by differentiation of primary basaltic melts; (2) primary crustal melts or derivative melts formed by crustal contamination of primary basaltic mantle melts; and (3) melts, or their derivatives, formed from either the crustal or the mantle parts of lithosphere descending along inclined seismic zones.

The mean $^{87}Sr/^{86}Sr$ ratio of the andesitic suites from the Marianas, New Britain, Izu Islands, Cascade Range, and Central America listed by DICKINSON is 0.7037 ± 0.0003. This is identical to the mean of 0.7037 ± 0.0001 that we calculated earlier for oceanic-island basalts. If these andesitic magmas had been significantly contaminated with older sial, their $^{87}Sr/^{86}Sr$ ratios would necessarily have been higher. Thus the strontium isotope data eliminate all but very minor sialic contamination for the andesites listed above. On the other hand, the $^{87}Sr/^{86}Sr$ ratios of andesites from New Zealand average about 0.7055 (EWART and STIPP, 1968; DICKINSON, 1970) a figure high enough to permit some crustal contamination to have occurred. Fig. IV.1 shows that the ratios of andesites are very similar to those of island basalts, even when the andesitic rocks are relatively poor in strontium content. DICKINSON concludes, "On balance, the available data on strontium isotopes afford no general support for the contamination or the crustal-fusion hypothesis . . .". He and many other geologists believe that andesites form by melting along arc-trench subduction zones (category 3 outlined above).

e) Antarctic and Tasmanian Dolerites

HEIER et al. (1965) reported isotopic and geochemical data for a number of specimens of tholeiitic dolerite of Jurassic age from Tasmania, rocks that crop out over an area of more than 6000 square miles. COMPSTON et al. (1968) presented similar data for the Jurassic Ferrar dolerite of Antarctica, which can be traced laterally for more than 1000 miles in Victoria Land. In the reconstruction of the continents by SMITH and HALLAM (1970), these two outcrop areas are adjacent. FAURE et al. (1968a, b, 1970) have also reported $^{87}Sr/^{86}Sr$ ratios of Antarctic basaltic rocks. All of these authors found that the Jurassic dolerites from Tasmania and from the Transantarctic Mountains — rocks that occur in enormous volumes — have ratios of K/Rb, U/K, Th/K, and initial $^{87}Sr/^{86}Sr$ that more closely resemble those of sialic rocks than those of basalts. Fig. IV.1 illustrates the $^{87}Sr/^{86}Sr$ ratios of these rocks; they range from about 0.708 to 0.718, and average 0.712.

COMPSTON et al. (1968) found the ratios of Karroo dolerites from South Africa and the Serra Geral dolerites from South America to average around 0.707, a value that is still high relative to oceanic basalts; MANTON (1968) also reported relatively high ratios in some Karroo basalts. Thus Mesozoic tholeiitic basalts from several Gondwanaland continents have $^{87}Sr/^{86}Sr$ ratios that are distinctly higher than those of oceanic basalts.

HEIER et al. (1965) and COMPSTON et al. (1968) considered several possible explanations of the high $^{87}Sr/^{86}Sr$ ratios of the Antarctic and Tasmanian dolerites, including (1) magma generation within the crust; (2) magma generation in regions of the mantle abnormally rich in Rb, U, and Th; (3) large-scale contamination of a mantle-derived basalt magma; (4) selective diffusion of certain elements into the magma; and (5) melting of sialic rocks by a hot basaltic magma and subsequent mixing of the two liquids.

The geophysical and petrologic evidence, which they summarize, seems to support a mantle origin, but if the dolerites come from an anomalous area of the mantle, it would have to have been of very large size. On the other hand, the fact that the rocks over a very wide area have high ratios appears to eliminate minor local contamination as a possible explanation. If contamination is the cause, it would have to have occurred on a large scale and prior to the emplacement of the rocks. HEIER et al. and COMPSTON et al. show that the addition of 10 to 30 percent of average crust to average oceanic tholeiite would produce rocks with the chemical and isotopic features of the dolerites, as would the addition of somewhat smaller amounts of material nearer to granite than to average crust in composition. But they reject the possibility of bulk assimilation, and prefer instead the hypothesis of selective diffusion of elements into the magma.

One distinguishing feature of the Tasmanian and Antarctic Jurassic dolerites is that their strontium contents average only about 130 ppm —

unusually low for continental basaltic rocks. As we stated earlier, the lower the strontium content of a magma, the more its $^{87}Sr/^{86}Sr$ ratio can be affected by a given degree of contamination. Thus the low strontium contents of the dolerites are consistent with the hypothesis that they were contaminated.

Some recent evidence bearing on the origin of the rocks comes from the work of FAURE et al. (1971). They analyzed specimens of the Ferrar dolerite, of the equivalent Kirkpatrick basalt, and of basalts and dolerites of probable

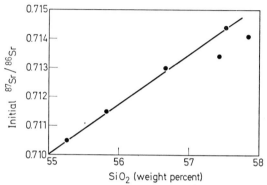

Fig. IV.4. The initial $^{87}Sr/^{86}Sr$ ratio versus the SiO_2 content in some basaltic rocks of Antarctica. The positive correlation suggests that the parent magmas of some of the basaltic rocks were contaminated with sialic material. (After FAURE et al., 1970)

Jurassic age from Queen Maud Land. Only the latter were found to have low $^{87}Sr/^{86}Sr$ ratios like those of oceanic basalts. Fig. IV.4 is a plot of initial $^{87}Sr/^{86}Sr$ ratio versus SiO_2 content, showing their positive correlation, in the Kirkpatrick basalts of Storm Peak. It is important to remember that such isotopic variations cannot be due to fractionation. Each of the six specimens of Kirkpatrick basalt comes from a different flow unit, and within the whole succession both $^{87}Sr/^{86}Sr$ ratio and SiO_2 content *decrease* from bottom to top. FAURE et al. concluded that the evidence "... suggests strongly that the chemical compositions of the Mesozoic basalts and dolerites of Antarctica have been affected by mixing of basalt magma with sialic material, presumably derived from the underlying Precambrian basement." Their work appears to offer a solution to the origin of the Antarctic dolerites, and — by implication — those from Tasmania, and their anomalous strontium isotopic ratios: the rocks have been contaminated on a large scale. It is interesting to speculate on whether the initiation of this large-scale mixing process may have been connected somehow with the breakup of Gondwanaland.

3. Variations in the Initial ^{87}Sr/^{86}Sr Ratio among and within Geologic Provinces

a) Oceanic Islands

Basaltic rocks from oceanic islands show a rather narrow range of initial ^{87}Sr/^{86}Sr ratio, as discussed above. When, however, intermediate and silicic rocks are included, significant isotopic variations are found among the rocks of neighboring islands, and even within individual islands. These variations may provide some useful constraints on the theories of formation of cogenetic volcanic rocks.

GAST et al. (1964) found that volcanic rocks from Gough Island in the Atlantic Ocean had significantly higher ratios than similar rocks from Ascension Island and that the ratios of rocks within each island were variable. They found analogous variations in the isotopic composition of lead from the same rocks, and concluded that the isotopic variations for both elements probably reflected differences inherent in the source regions of the rocks rather than in the effects of contamination. However, LESSING and CATANZARO (1964) found a significant inverse correlation between the ^{87}Sr/^{86}Sr and K/Rb ratios of rocks from the Hawaiian Islands (mainly from Hawaii), and suggested that this resulted from contamination of the parental magmas with pelagic sediments. POWELL et al. (1965) found that four undersaturated rocks from Oahu had significantly lower ^{87}Sr/^{86}Sr ratios than four more-silicic rocks from Hawaii. POWELL and DELONG (1966) analyzed several well-documented specimens from each of the three volcanic series on Oahu and found the mean ^{87}Sr/^{86}Sr ratio of the rocks of the tholeiitic Koolau series to be significantly higher than the mean for the undersaturated rocks of the Honolulu series, and the mean for the tholeiitic Waianae series. HEDGE (1966) reported what seem to be significant variations among the ^{87}Sr/^{86}Sr ratios of volcanic rocks from the island of St. Helena.

Thus at the present time there appear to be well-documented within-island variations in initial ^{87}Sr/^{86}Sr ratio for Ascension, Gough, St. Helena, Oahu, and Hawaii. The differences that have been observed are significant, but are quite small, and could have been lost in the scatter of interlaboratory precision had several laboratories been involved in the analyses for each island. It is entirely possible, for example, that small within-island variations of the type described by POWELL and DELONG (1966) for the different volcanic series of Oahu are the rule for volcanic islands, but that the sampling and replication of measurement have not been adequate to reveal them.

When within-island variations occur, the more silicic rocks usually have the higher ^{87}Sr/^{86}Sr ratios. On Ascension, Gough, and St. Helena these rocks are trachytes or phonolites; on Oahu they are tholeiites. On the other hand, in many instances trachytes, phonolites, and tholeiites do not have ratios that are demonstrably higher than those of associated alkali basalts, although this may in part reflect insufficient precision of measurement. At present it does not appear that the $^{87}Sr/^{86}Sr$ ratios of normal tholeiitic basalts from oceanic islands differ significantly and consistently from those of associated alkali basalts, although exceptions are known and both types from islands have higher ratios than the sea-floor basalts. It is certainly clear that if strontium isotopic differences between island tholeiites and alkali basalts exist, they are quite small and could be confirmed only by a series of very precise analyses. The continual improvement in the precision of measurement of $^{87}Sr/^{86}Sr$ ratios, as shown most dramatically by PAPANASTASSIOU and WASSERBURG (1969), suggests that this problem might well be re-opened.

b) Continental Volcanic Rocks

As illustrated in Fig. IV.1, the $^{87}Sr/^{86}Sr$ ratios of continental volcanic rocks, especially felsic ones, may be distinctly higher than those of oceanic volcanic rocks. As suggested above, these relatively high ratios could be primary and inherited from the source regions of the rocks, or they could be caused by contamination. In some cases, high and low initial $^{87}Sr/^{86}Sr$ ratios occur within a single igneous complex.

The Buck Hill and Tascotal Mesa Quadrangles of western Texas provide one interesting example of strontium isotope variations within an apparently cogenetic suite of continental volcanic and plutonic rocks. The volcanic and shallow-intrusive rocks of these two quadrangles include gabbro, basalt, andesite, trachyandesite, syenite, trachyte, and rhyolite, and were described by GOLDICH and ELMS (1949) and ERICKSON (1953). HEDGE (1966) analyzed 12 specimens from the area for $^{87}Sr/^{86}Sr$ ratio and rubidium and strontium contents. He expressed $^{87}Sr/^{86}Sr$ ratios relative to a standard, using $\Delta = (^{87}Sr/^{86}Sr$ unknown $- {}^{87}Sr/^{86}Sr$ standard$) \times 1000$.

HEDGE noted that in several of the continental suites that he analyzed, including the one from western Texas, the strontium in the felsic rocks was significantly more radiogenic than in the mafic ones. In order to explain this observation he favored a model of magma generation and differentiation at depth, followed by ascent and contamination of the derivative magmas at high levels in the crust.

To help evaluate the role of contamination in the genesis of these rocks we show graphs of $^{87}Sr/^{86}Sr$ Δ—values versus strontium content (Fig. IV.5) and versus silica content (Fig. IV.6). The $^{87}Sr/^{86}Sr$ and strontium data are from HEDGE (1966), and the silica contents are from the chemical analyses listed by GOLDICH and ELMS (1949) and ERICKSON (1953) for the rock types that Hedge analyzed. The graphs show that the three rocks with highest $^{87}Sr/^{86}Sr$ ratios, two rhyolites (r) and a trachyte (t), have low strontium contents, which would have made their $^{87}Sr/^{86}Sr$ ratios more easily

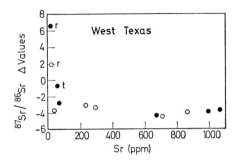

Fig. IV.5. A plot of initial $^{87}Sr/^{86}Sr$ $\varDelta$ — values versus strontium content for the igneous rocks of the Buck Hill (solid circles) and Tascotal Mesa (open circles) quadrangles, Western Texas. The data are from HEDGE (1966), who defined $\varDelta = (^{87}Sr/^{86}Sr$ unknown minus $^{87}Sr/^{86}Sr$ standard$) \times 1000$. The two samples with the highest $\varDelta$ — values are rhyolites (symbol r); the one with the next highest $\varDelta$ — value is a trachyte (symbol t). The $\varDelta$-values of the other samples do not differ significantly. The fact that the samples highest in $^{87}Sr/^{86}Sr$ ratio are very low in total strontium content is consistent with the hypothesis that they were contaminated with sialic material

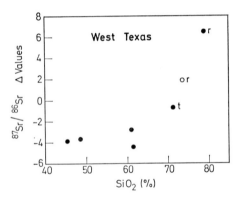

Fig. IV.6. A plot of initial $^{87}Sr/^{86}Sr$ $\varDelta$-values versus percent silica in the igneous rocks of the Buck Hill and Tascotal Mesa quadrangles, Western Texas. Silica contents are taken from GOLDICH and ELMS (1949) and ERICKSON (1953). In some cases they are averages of two or three analyses. See the caption for Fig. IV.5 for an explanation of the symbols. The correlation between the two parameters in the trachyte and rhyolites also supports the contamination hypothesis

changed by contamination. They are also highest in silica content, consistent with the suggestion that they have suffered additions of sialic material. An alternative hypothesis is that the trachyte and rhyolites were produced by partial melting of crustal rocks, the necessary heat having been supplied by the intrusion of the hot basaltic magma.

4. Possible Causes of Variations in Isotopic Composition among Cogenetic Rocks

In this chapter we have cited a number of examples of suites of apparently cogenetic volcanic rocks that show significant variations in initial $^{87}Sr/^{86}Sr$ ratio, and others have been reported. We close this chapter with a brief review and summary of the possible causes of such isotopic variations. As was the case with variations in isotopic composition among groups of volcanic rocks, two different explanations are possible. The first one is that the isotopic variations reflect differences in the Rb/Sr and $^{87}Sr/^{86}Sr$ ratios of the source regions of magmas. These differences could theoretically occur in several ways.

(1) The average Rb/Sr ratio of the upper mantle may decrease with depth — the $^{87}Sr/^{86}Sr$ ratio of a magma will then depend on its depth of origin.

(2) Individual mineral grains in the mantle may remain isotopically closed for long periods of time and develop different $^{87}Sr/^{86}Sr$ ratios. Partial melting of varying proportions of these minerals would then produce melts with different $^{87}Sr/^{86}Sr$ ratios. This hypothesis is not attractive since it implies that while one mineral is melting a nearby one is retaining much of its radiogenic strontium — an implication that contradicts the experience of geochronologists that radiogenic strontium in minerals is relatively mobile, especially at high temperatures.

(3) The Rb/Sr ratio of the mantle may vary laterally, and magmas may come first from one subcrustal site and then another. This could be brought about by migration of the site of melting underneath a static island, by a mobile island drifting across a static mantle, or by some combination of the two.

(4) ARTEMOV and YAROSHEVSKIY (1965) proposed that as a result of varying Rb/Sr ratios in different portions of a long-lived magma chamber, different $^{87}Sr/^{86}Sr$ ratios would be built up. They suggested that melts coming from different places within such a chamber would have variable $^{87}Sr/^{86}Sr$ ratios. It has usually been assumed that normal convective processes in magmas ensure strontium isotopic homogenization. However, NOBLE and HEDGE (1969) reported strontium isotope variations within individual ash-flow sheets, and concluded that the parent magmas of the ash flows became isotopically zoned because of contamination prior to eruption. Thus there is evidence that magmas need not always be homogeneous with respect to strontium isotopes.

(5) A related hypothesis (DICKINSON et al., 1969; Cox et al., 1970) is that a fractionated magma freezes with different Rb/Sr ratios at different levels within its chamber. If these different levels would be later remelted

and brought to the surface without mixing, isotopic variations would be observed in a suite of apparently cogenetic volcanic rocks. This hypothesis avoids the question of whether magmas are homogeneous with respect to $^{87}Sr/^{86}Sr$ ratio by having the isotopic differences develop after the magma has solidified but prior to remelting.

Neither hypothesis (4) nor hypothesis (5) imply that the ultimate source region, possibly the mantle, of cogenetic volcanic rocks is heterogeneous with respect to $^{87}Sr/^{86}Sr$ ratio. But, according to hypotheses (1), (2), and (3), spatially-related igneous rocks that have different isotopic compositions have come from different parent magmas that themselves were derived from different sources. In these cases, both isotopic and chemical variations among spatially-related volcanic rocks would be inherited from their source materials and not caused by fractional crystallization of a single parent magma.

The second general explanation is that the isotopic variations were produced by contamination. Several different mechanisms by which magmas could become contaminated with foreign strontium have been suggested.

(1) Bulk assimilation. One can calculate the amount of material — average crust for example — that would have to have been assimilated by a particular magma in order to produce a particular derivative rock. Such calculations often require excessive amounts of assimilation (PUSHKAR, 1967). Although bulk assimilation may remain a plausible mechanism in certain cases, models in which strontium is extracted preferentially by a magma from its wall-rock or from xenoliths may be more applicable.

(2) Wall-rock reaction. GREEN and RINGWOOD (1967) proposed that "incompatible elements," such as K, Rb, Ba, Sr, which do not enter the major mineral phases that are stable in the mantle, may be selectively transferred from the surrounding wall-rock into a magma. They suggest that at lower pressures within the crust, where plagioclase is a stable phase, strontium would be able to substitute for calcium and would no longer behave as an incompatible element.

(3) Selective migration of radiogenic strontium. HEIER (1964) and AL-RAWI and CARMICHAEL (1967) proposed that the radiogenic strontium in rubidium sites in mica and potassium feldspar would be more mobile than common strontium and would move more easily into an adjacent magma. This process might be an important part of wall-rock reaction, although some experiments, for example BAADSGAARD and VAN BREEMEN (1970), do not support its occurrence in nature.

(4) Isotope equilibration. PANKHURST (1969) found anomalously high $^{87}Sr/^{86}Sr$ ratios up to 0.712 (see Chapter VII) in the plutonic basic rocks of northeast Scotland and suggested that they had been produced by isotopic exchange or equilibration between a hydrous magma and the country rock.

This process could be particularly important in determining the $^{87}Sr/^{86}Sr$ ratios of strontium-poor ultramafic rocks, and it will be discussed at greater length in Chapter VII.

It has not been possible in most cases to decide between the alternatives of inhomogeneous source materials or contamination as the cause of observed isotopic variations, and indeed both may have operated for a particular suite. The greatest progress seems to have been made when both lead and strontium isotopic data were available (GAST et al., 1964), or where the geological and/or geochemical control were particularly good (PANKHURST, 1969; FAURE et al., 1970). It seems likely that combined lead and strontium isotopic studies on carefully-selected and chemically-analyzed volcanic rocks will lead to a better understanding of the cause of these variations. Such an understanding will be bound to have significant implications for the theory of petrogenesis of volcanic rocks.

5. Summary

The mean initial $^{87}Sr/^{86}Sr$ ratio of 164 basaltic rocks from oceanic islands is 0.7037 ± 0.0001 $(\bar{\sigma})$. The low-potassium sea-floor basalts often tend to have lower ratios than the island basalts, and PETERMAN and HEDGE (1971) showed that in fact a significant positive correlation exists between initial $^{87}Sr/^{86}Sr$ ratio and relative potassium content in oceanic basaltic rocks. The existence of these isotopic variations suggests that the mantle is not completely mixed, even on a time scale of hundreds of millions or billions of years.

The mean initial $^{87}Sr/^{86}Sr$ ratio of Circum-Pacific andesites from several different areas is 0.7037 ± 0.0003 — identical to the figure for island basalts. Only andesites from New Zealand seem to have ratios (average $= 0.7055$) significantly higher than those of island basalts. Thus most andesites that have been analyzed so far for initial $^{87}Sr/^{86}Sr$ ratios can have incorporated only minor amounts of older sialic material.

The Jurassic dolerites of Antarctica and Tasmania have puzzled geologists because their ratios of K/Rb, U/K, and initial $^{87}Sr/^{86}Sr$ (average about 0.712) are typical of those of sialic rocks. However, FAURE et al. (1970) found a very significant positive correlation between initial $^{87}Sr/^{86}Sr$ ratio and silica content in a suite of Jurassic dolerites from Antarctica, strongly suggesting that the high ratios of these rocks — and by implication those from Tasmania — were caused by contamination.

An increasing number of apparently-cogenetic suites of both oceanic and continental volcanic rocks are being discovered to have significant within-suite variations in initial $^{87}Sr/^{86}Sr$ ratio. In general, these variations could have been caused either by (1) differences in the initial $^{87}Sr/^{86}Sr$ ratios of the source regions of the rocks, or (2) variable contamination of their parent magmas with foreign strontium. It is important to discover which of the two is the cause in specific cases for the following reason: If two apparently cogenetic volcanic rocks have significantly different strontium isotopic ratios and can be shown not to have been contaminated, then they may not be related by fractional crystallization. If they are not, then some other process, perhaps partial melting, was the dominant mechanism by which they were generated.

V. Granitic Rocks

1. Introduction

Granites are coarse-grained crystalline rocks of igneous aspect composed of quartz, feldspar, and micas or ferromagnesian minerals such as hornblende. In this chapter we shall include under the heading of granitic rocks such related varieties as quartz monzonite, granodiorite, quartz diorite, and monzonite — all of which are mineralogically similar to granites. These granitic rocks collectively make up the large batholiths of the world, and many of the smaller stocks, laccoliths, dikes, and sills. They are by far the most abundant plutonic igneous rocks in the continental crust. If the scope of the definition were widened to encompass gneissic rocks of similar mineral composition, which may have had granites as parents, many of the rocks of the Precambrian Shield areas would also be included.

Granitic rocks occurring in the major batholiths of the world, such as the Sierra Nevada Batholith of California, have been studied by many geologists, and several conflicting hypotheses regarding their origin have been proposed. The ensuing debate has generated a voluminous literature. The opposing viewpoints have been summarized in works by GILLULY (1948), RAGUIN (1965), READ (1957), TURNER and VERHOOGEN (1960), and TUTTLE and BOWEN (1958), among others.

2. The Granite Problem

The crux of the "granite problem" is that it is possible that granitic rocks of igneous aspect can form as products of different processes, such as differentiation of basaltic parent magma, partial melting of rocks of different composition in the crust, and metasomatic reactions involving recrystallization and diffusion of ions on a large scale. If granitic rocks have formed by all of these and related mechanisms, the problem is to develop criteria by which the products of each different process can be distinguished. It has not been easy to find such criteria, because granitic rocks formed by each process could be very similar in bulk chemical composition and in mineralogy. However, an important aspect of the granite problem is the origin and prior history of the chemical elements which compose granitic rocks: Before these elements were combined to form granite, did they reside in the mantle or in the sialic crust? Our discussion in Chapter III suggested how the initial $^{87}Sr/^{86}Sr$ ratios of granitic rocks could shed light on this question. To summarize: If a granite has been derived by fractional crystallization of basalt

magma or by partial melting of deep crust or upper mantle, its initial
$^{87}Sr/^{86}Sr$ ratio should be low, like the ratios of most basalts. If, on the other
hand, the granite has been derived from older sialic material, the granite
should have inherited, as *its* initial $^{87}Sr/^{86}Sr$ ratio, the relatively higher ratio
that the older sialic material had at the time the granite was produced.

Thus low $^{87}Sr/^{86}Sr$ ratios preclude the possibility that relatively large
amounts of older sialic material participated in the formation of a granite.
On the other hand, high initial ratios do not prove that a granite *was* formed
solely by reconstitution of pre-existing sialic rocks, because high ratios may
also result from sialic contamination of a granite magma of mantle or lower-
crustal origin, or from a metamorphic event that caused an originally
magmatic granite to undergo isotopic equilibration on a whole-rock scale
(see Chapter IX). But in either of the latter two cases we may still conclude
that the granite contains material derived from a sialic system that is older
than the age given by the whole-rock Rb-Sr isochron for the granite — in
other words, that it had a crustal pre-history. This would be a very important
conclusion if it were found to apply to many granitic rocks.

3. Observed Initial $^{87}Sr/^{86}Sr$ Ratios of Granitic Rocks

The results of a survey of the literature, completed August 1971, are
listed in the Appendix. Since almost all of the initial ratios in the table come
from whole-rock isochron studies, several times as many individual granite
specimens were analyzed as suites are listed, and well over 500 specimens are
represented. We believe that enough data are available so that broad conclusions drawn now will not be significantly affected by the inclusion of data
obtained in the future.

The following criteria were used in selecting the data listed in the Appendix.
(1) Only rocks with granitic compositions and none with textures sufficiently metamorphic in appearance to allow them to be called granite gneiss,
gneissic granite, etc., are listed. (2) All of the initial $^{87}Sr/^{86}Sr$ ratios were
obtained from whole-rock analyses. No initial ratios obtained from combined whole-rock and mineral isochrons were used. Almost all of the initial
ratios are isochron intercepts, but as noted in the table, a few either were
measured directly or were calculated from the present-day $^{87}Sr/^{86}Sr$ ratio
and known age of a specimen. The age of the oldest rock for which this was
done is 60 million years.

Both of these criteria were adopted in order to avoid inclusion of samples
whose initial $^{87}Sr/^{86}Sr$ ratios were likely to have been affected by metamorphic re-equilibration of the strontium isotopes. Therefore, Appendix A
contains those analyzed granitic rocks that are most likely to have crystallized from a magma. The data are also presented in Fig. V.1 as a graph of
initial $^{87}Sr/^{86}Sr$ ratio against age.

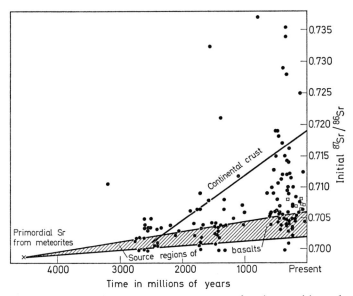

Fig. V.1. A plot of initial ^{87}Sr/^{86}Sr ratio versus age for the granitic rocks listed in the Appendix. With only a few exceptions, both age and initial ^{87}Sr/^{86}Sr ratio come from whole-rock rubidium-strontium isochron studies. The shaded zone delineates the approximate ^{87}Sr/^{86}Sr ratios of the source regions of oceanic basaltic rocks. It was outlined by drawing lines from the primordial meteorite point to the approximate maximum and minimum values observed in oceanic basalts, 0.706 and 0.702. The drawing of these as *straight* lines assumes that the Rb/Sr ratio of the source regions did not change with time; in effect, this conceives of the mantle as an "infinite reservoir" of rubidium and strontium. In fact, the removal of these two elements to the crust must have caused the Rb/Sr ratio of the mantle to change gradually, and probably to decrease, with time. In this case the boundary lines for the basalt zone shown should be convex upward. HART and BROCKS (1970) have discussed this question. See Fig. XII.2 for another interpretation of these data. The line labelled "continental crust" shows the development of the ^{87}Sr/^{86}Sr ratio in a 2500-million-year old system with an initial ratio similar to those of basalts and a Rb/Sr ratio of 0.18. The position of this development line is arbitrary, and could be changed by altering the initial ratio, or the Rb/Sr ratio, or the time of separation assumed. However, it is unlikely that the average age and Rb/Sr ratio of the continental crust are much greater than 2500 million years and 0.18, and therefore this development line probably sets an approximate upper limit to the average ^{87}Sr/^{86}Sr ratio of the continental crust at any point in time. The square boxes show the initial ^{87}Sr/^{86}Sr ratios and approximate ages of the major granitic batholiths of North America that have been analyzed for ^{87}Sr/^{86}Sr

As indicated in Section 2 above, by comparing the initial ^{87}Sr/^{86}Sr ratios of granites with those of basalts, we may be able to determine whether granites come from the same deep source regions as basalts or whether their formation involves significant amounts of older sialic material. However, many of the granitic rocks listed in the Appendix are Precambrian in age, and therefore in order to compare their initial $^{87}Sr/^{86}Sr$ ratios with those of basalts, it is necessary to take into account the fact that the $^{87}Sr/^{86}Sr$ ratios of the source regions of basalt magma have increased with time. We cannot compare the initial $^{87}Sr/^{86}Sr$ ratios of very old granites with those of modern basalts, or vice versa; the comparison must be between granites and basalts of roughly the same age. Although we do not know exactly how the $^{87}Sr/^{86}Sr$ ratio of the mantle has evolved (this is discussed in Chapter XII), we can approximate it as we did in Fig. III.1 by drawing lines from the maximum and minimum ratios presently observed for oceanic island basalts, approximately 0.706 and 0.702 (see Chapter IV), back to the primordial meteorite point of $(^{87}Sr/^{86}Sr)_0 = 0.699$; $t = 4600$ million years. Also shown in Fig. V.1 is the development line that we plotted in Fig. III.1 for a 2.5 billion-year-old system that originated with a basaltic initial $^{87}Sr/^{86}Sr$ ratio and a Rb/Sr ratio equaling 0.18. Fig. V.1 will form the basis for our discussion and interpretation of the initial $^{87}Sr/^{86}Sr$ ratios of granitic rocks.

GERLING et al. (1968) constructed a graph similar to Fig. V.1, using the ages and initial $^{87}Sr/^{86}Sr$ ratios of 54 granitic suites, including a number of those we have included in Appendix and Fig. V.1. They fitted a regression line to their data by rejecting points whose deviation from a direct regression exceeded three times the standard deviation. They obtained a linear correlation coefficient (r) of 0.60, and found that the regression line intersected at $^{87}Sr/^{86}Sr = 0.708$ for $t = 0$, and at $^{87}Sr/^{86}Sr = 0.700$ for $t = 4580$ million years. If the granites which they plot have come from the same general source region, then the slope of their regression line is proportional to the Rb/Sr ratio of that region and the line itself shows the growth of ^{87}Sr in the source region with time.

We have not attempted a similar regression of the data shown in Fig. V.1 because the wide range of initial ratios observed strongly suggests that these granitic rocks have had neither a common source nor simple one-stage histories. A regression line through a series of unrelated points has no significance. Nor does it appear to be justified to reject granites having anomalously high initial $^{87}Sr/^{86}Sr$ ratios. GERLING et al. listed five suites with initial ratios above 0.7120, and rejected three of them. We list 23 suites with initial ratios above 0.7120.

4. Interpretation and Discussion of Initial $^{87}Sr/^{86}Sr$ Ratios of Granites

a) Granites with Initial Ratios within the Basalt Field

About 50 percent of the points plotted in Fig. V.1 either fall in the zone we shall call the "basalt field," or lie no more than about 0.001 units of $^{87}Sr/^{86}Sr$ ratio above or below it. If continental basalts, especially the Antarctic and Tasmanian dolerites, had been used to define the basalt field, it obviously would have been wider, and more of the points would have fallen within it. But, as we explained in Chapter IV, we believe that the continental basaltic rocks with relatively high initial $^{87}Sr/^{86}Sr$ ratios have likely been contaminated with radiogenic strontium, and if so their ratios are not representative of those of the deep source regions of most oceanic and continental basalts. Oceanic basalts, on the other hand, cannot have been significantly contaminated with crustal radiogenic strontium, and their initial $^{87}Sr/^{86}Sr$ ratios should accurately reflect those of their source regions in the upper mantle. We assume that the $^{87}Sr/^{86}Sr$ ratios of the suboceanic mantle and the subcontinental mantle are not significantly different, and of course the basal part of the crust could have the same $^{87}Sr/^{86}Sr$ ratio as the upper mantle. In our analysis of Fig. V.1 we are in effect asking what percentage of the granites represented *could* have come from the same ultimate source region as oceanic basalts — the upper mantle — by partial melting of upper-mantle material or by differentiation of primary basalt magma. Our conclusion is that about 50 percent of the granitic-rock suites analyzed have initial $^{87}Sr/^{86}Sr$ ratios that are consistent with such an origin. To put it another way, one-half of the granitic rocks analyzed to date cannot contain more than very small amounts of older crustal strontium.

The Salisbury Pluton, North Carolina. As a specific example of a granitic rock having an initial $^{87}Sr/^{86}Sr$ ratio within the basalt range we have selected for discussion the Salisbury Pluton of North Carolina, whose petrography, and general and strontium isotopic geochemistry were described by FUL-LAGAR, LEMMON, and RAGLAND (1971). This granitic pluton is one of a number occurring in the Piedmont of the southeastern United States. An isochron diagram is shown as Fig. V.2.

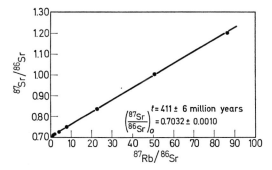

Fig. V.2. The Rb/Sr isochron diagram for the Salisbury Pluton of North Carolina. Its initial $^{87}Sr/^{86}Sr$ ratio is identical to those of many basalts, which indicates that its parent magma (1) originated at the base of the crust or in the upper mantle, and (2) incorporated little or no strontium from older crustal rocks. (After FUL-LAGAR, LEMMON, and RAGLAND, 1971)

FULLAGAR et al. were able to conclude from the major element geochemistry of the pluton that it had undergone "eutectic point" crystallization in the system Ab—Or—An—Q—H_2O and that it had been metasomatic-
ally altered. The latter conclusion raised the possibility that the age obtained
from the isochron was in fact an age of metasomatism and that the true,
original age of crystallization was greater. However, they pointed out that
for the data points on Fig. V.2 to have remained linear *after* metasomatism,
both rubidium and strontium would have to have been added or removed
by the same percent from every sample — a very unlikely possibility. On the
other hand, if metasomatism had occurred early in the history of the pluton,
while its isochron was almost horizontal (see Fig. II.2), changes in Rb/Sr
ratio would have simply caused the points to shift to the right or left, leaving the linearity of the isochron nearly unaffected. FULLAGAR et al. therefore
concluded that the pluton had undergone metasomatism very shortly after
it crystallized.

The initial ratio of the Salisbury granitic rocks, 0.7032 ± 0.0010, as
illustrated in Fig. V.1, lies almost squarely in the center of the basalt field.
FULLAGAR et al. wrote that this low initial $^{87}Sr/^{86}Sr$ ratio indicates that
". . . the original magma from which the Salisbury magma fractionated
originated in the mantle . . ." and ". . . incorporated little or no strontium
from older crustal rocks . . .".

b) Granites with High Initial $^{87}Sr/^{86}Sr$ Ratios

Approximately 20 percent of the granites represented in Fig. V.1 have
initial $^{87}Sr/^{86}Sr$ ratios that are higher than or that lie within 0.001 units of
$^{87}Sr/^{86}Sr$ ratio of the line for the continental crust. These high initial ratios
are unlikely to be the result of minor contamination, and we interpret them
to mean that a significant fraction of the material making up these granites
existed as sialic material before the granites themselves were formed.

The Heemskirk Granite, Western Tasmania. The Heemskirk granite of
Western Tasmania (BROOKS and COMPSTON, 1965; BROOKS, 1966, 1968) is
one of the most thoroughly-studied rock units in the world in terms of
rubidium and strontium isotopic measurements, and it provides an excellent
example of the application of strontium isotopes to the granite problem. An
isochron diagram for the Heemskirk granite is shown as Fig. V.3.

The granite intrudes slates known to be within the range Upper Silurian
to Middle Devonian in age, and thus the granite itself cannot be older than
about 400 million years. The initial $^{87}Sr/^{86}Sr$ ratios of the three phases of the
granite, as determined by the whole-rock measurements illustrated in
Fig. V.3, are 0.734, 0.741, and 0.719. The fit of the regression lines for
each of the white granite isochrons is excellent: all of the scatter about the
isochrons can be explained by the experimental precision (BROOKS and

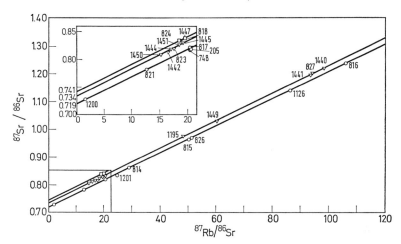

Fig. V.3. The Rb/Sr isochron diagram for whole-rock samples of the Heemskirk granite, Tasmania. Circle $\circ$ = red granite, triangle $\triangle$ white series A granite, square $\square$ = white series B granite. Two specimens (one red, one white B) were omitted by BROOKS and COMPSTON because of scale requirements. The high initial $^{87}Sr/^{86}Sr$ ratios shown could be due to post-crystallization processes such as isotopic homogenization or metasomatic addition of radiogenic strontium, but it is difficult to see how any post-crystallization process operating on a whole-rock scale could produce or preserve the remarkably high degree of colinearity shown by the isochron for the white series A granite, for example, BROOKS and COMPSTON therefore concluded that the granites were already enriched in ^{87}Sr at the time they formed. (After BROOKS and COMPSTON, 1965)

COMPSTON, 1965). The red granite and the white Series A granite give ages of 354 million years; the white Series B granite may be slightly older.

BROOKS and COMPSTON (1965) give a thorough discussion of the possible interpretations of the high initial ratios of the various phases of the Heemskirk granite, and our presentation follows theirs. First, it is possible that the granite formed at some time prior to 354 million years ago, developed higher $^{87}Sr/^{86}Sr$ ratios with time, and underwent a metamorphic event 354 million years ago which caused complete internal redistribution and homogenization of its strontium isotopes. LANPHERE et al. (1964) have described whole-rock systems that became opened to ^{87}Sr in this way. To test this suggestion we can assume that the 400-million-year maximum age is in fact the true age of emplacement for the Heemskirk granite, and use Eq. (II.15) to calculate what its initial $^{87}Sr/^{86}Sr$ ratio would have been at that time. BROOKS and COMPSTON (1965) used such a procedure to show that in the period from 400 to 354 million years ago, the average Rb/Sr ratio of the granites could have caused an average increase in their $^{87}Sr/^{86}Sr$ ratios of only 0.009. Thus if the granite were 400 million years old, the initial $^{87}Sr/^{86}Sr$ ratios of the two white granite phases, for example, would have been $0.734 - 0.009 = 0.725$, and $0.741 - 0.009 = 0.732$. Even if their true age is 400 million years and if 354 million years is an age of metamorphism, the granites would *still* be marked by abnormally high initial $^{87}Sr/^{86}Sr$ ratios. Therefore, internal redistribution and homogenization of the strontium isotopes during metamorphism is not a sufficient explanation of their high initial $^{87}Sr/^{86}Sr$ ratios. When the maximum geological age of a granite cannot be determined, which is unfortunately the situation for a number that have been analyzed by the whole-rock Rb-Sr method, such limits cannot be set, and isotopic homogenization on a whole-rock scale remains as a possible, if unlikely, explanation of high initial $^{87}Sr/^{86}Sr$ ratios.

BROOKS and COMPSTON (1965) also considered the possibility that the high initial ratios of the Heemskirk granite were produced by migration of strontium from the surrounding country rock into the granite during a later metamorphic-metasomatic event. WASSERBURG et al. (1964a) thought that the high and variable initial $^{87}Sr/^{86}Sr$ ratios of dikes of Pahrump diabase in the Panamint Mountains of California were due to metamorphic-metasomatic effects, and although the Heemskirk granite represents a comparatively large volume of material, this mechanism must be considered. The principal evidence against it is that (1) the sedimentary rocks around the granite, except for the contact aureole, are unmetamorphosed, and (2) the two white-granite isochrons show an exceptionally-high colinearity. As was the case with the Salisbury pluton, it is very difficult to see how a postcrystallization addition of strontium to a coarse-grained granite could possibly create or preserve the near-perfect fit observed for the two white-granite isochrons. As BROOKS and COMPSTON (1965) point out, this postulate also fails to explain the difference between the initial $^{87}Sr/^{86}Sr$ ratios of the two white granites, which are adjacent and should have been similarly metasomatized.

Thus the general hypotheses of post-crystallization addition and of internal homogenization of radiogenic ^{87}Sr fail, and one must conclude that the Heemskirk granite was already enriched in radiogenic strontium at the time it crystallized. A possible explanation of this enrichment is that the parent magma of the Heemskirk granite was derived entirely from older sialic material by granitization or by partial melting. Another possibility is that the parent magma was of deep origin, perhaps even coming from the upper mantle, and was contaminated with radiogenic strontium on its way to the surface. In either case the formation of the Heemskirk granite magma must have involved in part at least the reconstitution of significant amounts of older, presumably sialic, material.

c) Granites with Intermediate Initial $^{87}Sr/^{86}Sr$ Ratios

Now let us return our attention to Fig. V.1, and consider the remaining group — granites with initial $^{87}Sr/^{86}Sr$ ratios intermediate between those of oceanic island basalts and those of the limiting development line for average crust. Two types of granitic rocks are represented in this category: (1) smaller plutons and stocks; for example, the Cape granite of South Africa, the Llano granite of Texas, the granites of Skye, etc., and (2) the Phanerozoic batholiths of North America. Before discussing category (2), we shall consider briefly one example of category (1) — the strontium isotope data of MOORBATH and BELL (1965) for Skye.

The Isle of Skye. This island belongs to the British Tertiary volcanic province, and is a classic petrologic area (HARKER, 1904). It comprises three principal geologic units: a northern area of plateau basalts, a central Tertiary intrusive complex, and a southern upland of sedimentary and metamorphic rocks. The intrusive complex consists of a basic and ultrabasic layered intrusion and several granitic centers. According to Moorbath and Bell, the origin of the granitic rocks has been attributed by various authors to (1) fractional crystallization of basaltic magma, (2) metasomatism of pre-existing rocks, and (3) partial melting of country rock. MOORBATH and BELL point out that option (2) is unlikely since the granites show marked intrusive features and since volcanic rocks of granitic composition do exist on Skye.

The strontium isotope data of MOORBATH and BELL are shown as a histogram in Fig. V.4. They wrote (p. 55), "The lower mode is made up entirely of values obtained for basic and ultrabasic rocks. The upper mode consists solely of results from granites and from rocks of the marscoite [hybrid] suite." Their results show very clearly that the granitic rocks of Skye have not been formed simply by fractional crystallization of the parent

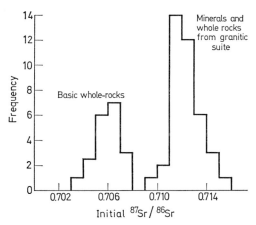

Fig. V.4. Histogram of the initial ^{87}Sr/^{86}Sr ratios of igneous rocks from the Isle of Skye. The distinctly-higher ratios of the granitic rocks show that they were not derived from the parent magma of the basic rocks by fractional crystallization, and suggest that the granitic rocks contain crustal strontium. (The data are from MOORBATH and BELL, 1965)

4*

magma of the basic and ultrabasic rocks; otherwise all would have had the same initial $^{87}Sr/^{86}Sr$ ratio. MOORBATH and BELL concluded that the granitic rocks probably were formed by partial melting of the gneisses of the Lewisian basement complex (option 3 above), which could have had appropriate $^{87}Sr/^{86}Sr$ ratios at the time of intrusion of the granites. They suggest that the heat for this melting was provided by the adjacent hot basic magma.

Phanerozoic Batholiths of North America. Although granites with initial $^{87}Sr/^{86}Sr$ ratios intermediate between the "basalt field" and the average crustal development line represent only about 30 percent of those plotted on Fig. V.1, the presence of all of the major Phanerozoic batholiths of North America that have been analyzed makes this group by far the most important in terms of the volume of granitic rock represented. Included are these batholiths: Nova Scotia [$(^{87}Sr/^{86}Sr)_0 = 0.708$], Sierra Nevada (0.7073), Coast Range of California (0.7082), Inyo Mountains of California (0.7070), British Columbia (0.7071), and Boulder (0.7066). These figures probably are within experimental error of each other, and a value of 0.707 ± 0.001 thus appears to be characteristic of the initial $^{87}Sr/^{86}Sr$ ratios of the North American Phanerozoic batholiths. An initial ratio of 0.707 is too low to have been produced by the melting or metasomatism of average crustal materials, and is just slightly higher than expected for the upper mantle, as illustrated on Fig. V.1. Below we list and evaluate the mechanisms that we believe could conceivably produce granitic rocks with these intermediate initial $^{87}Sr/^{86}Sr$ ratios; HAMILTON and MEYERS (1967) also discuss this subject. We do not include some mechanisms that have been proposed that we feel are unlikely to be significant.

(1) A melt of granitic composition is derived by differentiation in, or by partial melting of, the lower crust or upper mantle, and is subsequently contaminated with radiogenic strontium at high levels in the crust.

(2) Granitic liquids are produced by partial melting of crustal rocks with appropriate Rb/Sr ratios (see DOE, 1968).

(3) Mixtures of materials of mantle origin (with low $^{87}Sr/^{86}Sr$ ratios) and materials of intermediate or upper crustal origin (with somewhat higher $^{87}Sr/^{86}Sr$ ratios) undergo partial melting.

Contamination of a granitic melt at high levels in the crust (mechanism 1) must often occur if such melts exist, and this could explain the ratios of many smaller granitic masses with intermediate or high initial ratios. However, it seems unlikely that this process can explain the uniformity of the $^{87}Sr/^{86}Sr$ ratios of the North American batholiths listed above.

Mechanism 2, by its nature, is difficult to confirm or reject, though it is possible to develop an argument in favor of it. For example, we know that some ancient shield areas underwent granulite facies metamorphism 2500 to 3000 million years ago and that granulites have low Rb/Sr ratios (HEIER, 1964). If the basal part of the crust experienced such early granulite facies metamorphism, its low Rb/Sr ratio could have generated ^{87}Sr/^{86}Sr ratios similar to those observed in the batholithic granites.

The volcanic and sedimentary rocks found in eugeosynclinal sequences have appeared to many petrologists to be the likely parent materials of granitic batholiths (see TURNER and VERHOOGEN, 1960, pp. 382—388). PETERMAN et al. (1967) showed that the Rb/Sr and ^{87}Sr/^{86}Sr ratios of such assemblages are similar to those of many batholithic granites. Thus it seems that partial melting of eugeosynclinal volcanic and sedimentary rocks, one possible variation of mechanism 3, is a petrogenetic hypothesis that is consistent with both the isotopic and the geologic evidence for granitic batholiths.

Within the last few years, however, the whole concept of a geosyncline has begun to undergo serious re-evaluation (for example, DEWEY and BIRD, 1970). With this re-evaluation has come the proposal that eugeosynclinal rock sequences are in fact oceanic crust — sediment and volcanic rocks — that have been added to continents along subduction zones at continental margins. Thus the past existence of eugeosynclines, much less their partial melting to produce granites, is now under question.

The application of the theory of plate tectonics and subduction zones to isotopic data was carefully developed by ARMSTRONG (1968). He attempted to reconcile the lead isotope evidence for ancient continents (PATTERSON and TATSUMOTO, 1964) with the strontium isotope evidence of HURLEY et al. (1962) that the continents have continually received new additions of juvenile material. In ARMSTRONG's model, granitic material is eroded from the continents and carried into the oceanic basins, where it is caught on the spreading sea-floor. Eventually it is dragged down along a subduction zone to be remelted, mixed with mantle material, and finally emplaced once more in the crust. Thus the crust is in a state of cyclical evolution, and juvenile-appearing igneous rocks actually are composed in part of recycled older sial. The evolution of lead isotopes is dominated by mixing in the crust, that of the strontium isotopes by the vast reservoir of mantle strontium.

The implications of plate tectonics for continental geology have been developed by, among others, HAMILTON (1969a, b), who outlined a model for the evolution of the batholiths of North America, and by DEWEY and BIRD (1970), who described the relationships between plate movement and orogeny. The view of these geologists is that the material of the granitic batholiths originated in the mantle at and above the Benioff zone near a former continental margin. The mobile core from which these batholiths ultimately form possibly begins as a "minimum-melting mixture" of sedimentary, volcanic, and mantle materials. It starts with a relatively low ^{87}Sr/^{86}Sr ratio, and as it rises through the overlying mantle and crust, it may react with volcanic rocks and oceanic and terrigenous sediments. When finally emplaced, these batholiths contain a mixture of mantle and crustal strontium, but their ^{87}Sr/^{86}Sr ratios are only slightly higher than those of the upper mantle because (1) the mantle strontium is more abundant and dominates, and (2) the sedimentary and volcanic rocks with which they react are relatively young and, with the possible exception of some oceanic sediments, did not inherit and have not yet had time to develop large amounts of radiogenic strontium. The observation that the Phanerozoic batholiths of North America have $^{87}Sr/^{86}Sr$ ratios of about 0.707 $\pm$ 0.001 seems to be entirely consistent with the models of ARMSTRONG (1968), HAMILTON (1969a, b), and DEWEY and BIRD (1970), and it appears that this process can explain both the geologic and the isotopic features of granitic batholiths. At the present time it does not appear to be possible to choose between this variety of mechanism 3, and mechanism 2, as postulated above, as explanations of the petrogenesis of granitic batholiths.

6. Summary

The observed initial $^{87}Sr/^{86}Sr$ ratios of granites covering a wide span of ages show no preference for a narrow range of values, but scatter from about 0.700 to 0.737. About 50 percent of the granitic suites analyzed have ratios identical to those of oceanic basalts, and thus can contain only relatively minor amounts of older sialic material. Approximately 20 percent have ratios as high as or higher than an expected maximum for average continental crust — these rocks must contain relatively large amounts of recycled older crustal rocks.

The largest volume fraction of granitic rocks, including the analyzed major Phanerozoic batholiths of North America, have initial ratios of 0.707 $\pm$ 0.001. These intermediate initial $^{87}Sr/^{86}Sr$ ratios may indicate that the batholithic rocks are formed by partial melting of a basal crustal zone with the appropriate Rb/Sr ratio, or that they are composed of mixtures of basaltic and sialic materials. As an example of the latter, ARMSTRONG (1968), HAMILTON (1969a, b), and DEWEY and BIRD (1970), have shown how the material of granitic batholiths could originate by partial melting along subduction zones. These melts initially include mainly mantle material with some subducted oceanic crust, but as they rise they continually react with the overlying mantle and crust. As a result of reaction with crustal rocks, their $^{87}Sr/^{86}Sr$ ratios are gradually increased over their original, characteristically low, mantle values. Such a process could produce batholithic granites with observed initial $^{87}Sr/^{86}Sr$ ratios of about 0.707 $\pm$ 0.001.

VI. Alkalic Rocks and Carbonatites

1. Introduction

Alkalic rocks and carbonatites are among the rarest igneous rock-types. They tend to occur in stable platform areas and in Precambrian shields, rather than in orogenic belts. BARKER (1969) estimated that in North America, alkalic rocks occupy a total surface area of only about 1500 square miles, and HEINRICH (1966) concluded that the *worldwide* areal exposure of carbonatites is only about 200 square miles. Yet DALY (1933) wrote that as many as one-half of the names that have been given to igneous rocks refer to alkalic types. The literature of igneous petrology also reflects the interest which these rocks have held: one complete volume on alkalic rocks (SORENSEN, 1972) and two volumes on carbonatites (HEINRICH, 1966; TUTTLE and GITTINS, 1966) are available, and journal articles are numerous. HAMILTON (1968) and POWELL and BELL (1972) have reviewed the application of strontium isotope studies to the origin of alkalic rocks.

Alkalic igneous rocks have enough of the alkali elements to give them a distinctive mineralogy. They may contain a sodic pyroxene or sodic amphibole, or a feldspathoid such as nepheline or leucite. Syenite and nepheline syenite are perhaps the most common varieties, but many others occur, and a formidable nomenclature has developed.

Some syenites occur adjacent to granite, and the fine-grained equivalents of syenite and nepheline-syenite (trachyte and phonolite) occur on some oceanic islands. However, few $^{87}Sr/^{86}Sr$ ratio measurements are available for these two occurrences, and therefore we shall concentrate instead on continental alkalic rocks that are not associated with granite. We also exclude discussion of the lamprophyres as a separate group. Although some of them are alkalic, few have been analyzed for $^{87}Sr/^{86}Sr$ ratio.

Most of the hypotheses that have been proposed to account for the origin of alkalic rocks appeal to one of three mechanisms (see BELL and POWELL, 1969c): (1) Fractional crystallization, usually of basaltic or peridotitic magmas. (2) Assimilation or mixing. According to different authors, the materials mixed may be basalt magma and sialic rock, or granite magma and limestone, or carbonatite magma and sialic rock, etc. In a few cases mixing through metasomatism rather than through assimilation has been proposed. (3) Partial melting. Substances melted may be biotites and hornblendes, a zone deep in the crust, an old kimberlitic layer, etc. Zone melting (refining) also belongs in this group.

It may be possible using strontium isotopes to decide in which of these three broad categories certain classes of alkalic rocks belong. For example, rocks formed by fractional crystallization of mafic or ultramafic magmas (group 1) should have low $^{87}Sr/^{86}Sr$ ratios. Rocks formed by mixing of primary magmas and crustal materials (group 2) should have intermediate $^{87}Sr/^{86}Sr$ ratios and might show some patterns characteristic of mixing. These might include a positive, linear trend on a plot of $^{87}Sr/^{86}Sr$ versus Rb/Sr ratio, and an inverse, hyperbolic pattern on a graph of $^{87}Sr/^{86}Sr$ ratio versus strontium content. Rocks formed by partial melting of sialic materials (group 3) should have high initial ratios, whereas those formed by zone-melting of the upper mantle should have low and very uniform ratios.

2. Carbonatites

Carbonatites are carbonate (mainly calcite) rocks that appear to be igneous. HEINRICH (1966) estimates that over 300 individual occurrences are known. Carbonatites are almost always intimately associated with alkalic igneous rocks, which is the reason for discussing both types in a single chapter. The most common associates of carbonatite probably are ijolite (nepheline plus pyroxene) and nepheline syenite, although on occasion kimberlite joins the group. Many alkalic rocks (about 70 percent according to HEINRICH (1966)) occur without accompanying carbonatites, but most petrologists are nevertheless convinced that the two are closely related.

Carbonatites are highly enriched in certain trace elements, notably strontium, barium, and the rare earths. Their trace-element contents are distinctly different from those of limestones, although for many years carbonate rocks associated with alkalic rocks were believed to be limestone xenoliths whose presence supported the limestone syntexis hypothesis. However, carbonatites are known both in oceanic settings (Cape Verde Islands and Canary Islands) and as lava flows and pyroclastic deposits. There seems little doubt that most carbonatites have cooled from a melt. The carbonatite problem is now one of discovering the ultimate source of their CO_2 and the nature of their genetic relationship to alkalic rocks.

Around the margins of many carbonatite-alkalic rock complexes occurs a zone of metasomatized country-rock. The rocks in such zones are called fenites. Some petrologists, including VON ECKERMANN (1948) and a number of Russian authors, believe that *all* of the silicate rocks in carbonatite-alkalic rock complexes are metasomatic (possibly remobilized) and that only the carbonatite magma is primary. According to this view ijolites, for example, are thoroughly metasomatized rocks, or ultrafenites. Other petrologists, probably a majority, believe that the original magma was silicate, that most if not all of the rocks are magmatic, and that the carbonatite represents a final, residual differentiate of the parental silicate magma. HEIN-

RICH (1966) has likened this to the problem of the chicken versus the egg. WYLLIE (1966) reviewed the extensive work done on the physical chemistry of carbonatic liquids.

If the carbonatite and the silicate rocks within a complex are comagmatic, they should have the same initial $^{87}Sr/^{86}Sr$ ratio. On the other hand, if all of the silicate rocks are metasomatized country rock, some of them should have partly preserved the $^{87}Sr/^{86}Sr$ ratios of the country rock, and they should show patterns of $^{87}Sr/^{86}Sr$ ratio and elemental abundance that are characteristic of mixing. These patterns might be especially pronounced because of the contrast between the low $^{87}Sr/^{86}Sr$ ratios of carbonatites and the much higher ratios of the Precambrian gneisses and granites that surround many complexes. A potential difficulty, however, is that carbonatites are so enriched in strontium that in any mixing process carbonatitic strontium might mask strontium contributed from another source.

A histogram plot of the observed initial $^{87}Sr/^{86}Sr$ ratios of carbonatites is shown in Fig. VI.1. The diagram also includes data for sedimentary carbonate rocks; for carbonate "vein-dikes" (rocks that have some properties intermediate between those of carbonatites and those of hydrothermal veins); and for some skarn deposits (metamorphosed and metasomatized limestones) from the Grenville structural province of eastern Canada.

The histogram plot shows that as a group carbonatites have $^{87}Sr/^{86}Sr$ ratios that are distinctly lower than those of limestones, although there is overlap between the two. The ratios of the vein-dikes, which fall toward the upper end of the carbonatite range, are consistent with the hypothesis that the carbonatites and vein-dikes are genetically related. POWELL (1965) concluded that the vein-dikes had significantly higher ratios than those of carbonatites, but this conclusion is not supported by the greater amount of data now available. The skarns also have ratios that overlap those of carbonatites. GITTINS et al. (1969) wrote that the low $^{87}Sr/^{86}Sr$ ratios of some of the Grenville skarns are either primary features, in which case the ratios of non-igneous carbonate rocks overlap those of carbonatites, or they were produced by combined metamorphism and metasomatism. In either case, they implied, the $^{87}Sr/^{86}Sr$ ratios of carbonate rocks cannot be used to distinguish carbonatites from other types. The very old Bulawayan limestone, first analyzed by GAST (1955), with its low $^{87}Sr/^{86}Sr$ ratio of about 0.702, shows clearly that non-igneous carbonate rocks can have ratios as low as or lower than those of carbonatites.

SHAW et al. (1963) have shown that the Grenville skarns contain as much as 6600 ppm strontium and large amounts of other trace elements — barium and yttrium, for example. If rocks with such high trace-element concentrations are metasomatic, very extensive metasomatism of a parent rock must have occurred. If the parent in the case of the Grenville skarns was indeed

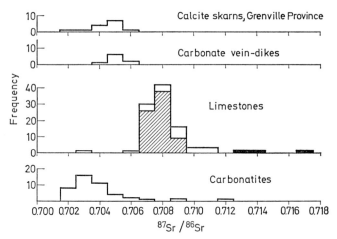

Fig. VI.1. Histogram of the initial $^{87}Sr/^{86}Sr$ ratios of carbonatites, limestones, carbonate vein-dikes, and calcite from skarns in the Grenville province of Canada. The mean $^{87}Sr/^{86}Sr$ ratio of the 42 carbonatites is 0.7034 ± 0.0003 ($\bar{\sigma}$). A small amount of overlap occurs between the carbonatite and limestone ranges, even if the Bulawayan limestone ($^{87}Sr/^{86}Sr = 0.703$) is excluded. The ratios of the vein-dikes and the skarns fall in the upper part of the carbonatite range. The histogram shows that the $^{87}Sr/^{86}Sr$ ratio of a carbonate rock is a good but imperfect predictor of whether the rock is a carbonatite or a limestone, and that carbonatites, carbonate vein-dikes, and some skarns cannot be distinguished unequivocally on the basis of $^{87}Sr/^{86}Sr$ ratio alone. Sources of data are: *Carbonatites*: BROOKINS (1967); BROOKINS and WATSON (1969); DEANS and POWELL (1968); POWELL (1966); and unpublished data of the authors. The data of HAMILTON and DEANS (quoted in BROWN, 1964) for two specimens of the Songwe scarp carbonatite are included in the figure (but not in the calculation of the mean), even though there may be some question as to whether it represents a bona fide carbonatite. The ratios that they obtained, 0.709 and 0.712, fall well outside the range shown by other carbonatites that have been analyzed. *Limestones*: PETERMAN et al. (1970): cross-ruled, analyses of individual Phanerozoic fossil specimens. HOEFS and WEDEPOHL (1968): solid bar, composite samples of (from lowest to highest $^{87}Sr/^{86}Sr$ ratio) 45 Jurassic, 16 Cretaceous, and 32 Devonian limestones. Also POWELL (1965c) and some of the references used for carbonatite analyses. *Vein-dikes*: POWELL (1965a). *Skarns*: GITTINS et al. (1969), and unpublished data of the authors

Grenville marble, the metasomatizing fluid (1) was very rich in strontium, barium, yttrium, and other trace elements, and (2) had a very low $^{87}Sr/^{86}Sr$ ratio. What was the nature and origin of such fluids? HEINRICH (1966, p. 311) has described the region of Quebec from which the skarns come as a "migmatitic carbonatite province," implying that the skarns in some cases may actually be carbonatitic rocks. Thus one could argue that the low $^{87}Sr/^{86}Sr$ ratios of the Grenville skarns, instead of negating the use of strontium isotopes for recognition of carbonatite, simply confirm the carbonatitic affinities of the skarns.

The histogram shows that of 43 carbonate rocks with $^{87}Sr/^{86}Sr$ ratios less than or equal to 0.704, 35 are carbonatites as defined by other criteria and only one is a sedimentary carbonate rock. Thus a sound empirical conclusion is that an initial $^{87}Sr/^{86}Sr$ ratio of 0.704 or less *suggests* that a carbonate rock for which the other geological and geochemical evidence is inconclusive is not sedimentary, and is probably a carbonatite. But it is clear that low $^{87}Sr/^{86}Sr$ ratios are not a conclusive criterion for recognition of typical carbonatites and that, conversely, ratios as high as 0.706 to 0.707 may not prove that a rock is not a carbonatite.

The mean $^{87}Sr/^{86}Sr$ ratio of the carbonatites shown, excluding the two from Songwe scarp, is 0.7034 ± 0.0003 $(\bar{\sigma})$. This figure is identical within its precision to the mean $^{87}Sr/^{86}Sr$ ratio of oceanic basaltic rocks, 0.7037 ± 0.0001 $(\bar{\sigma})$ (see Chapter IV). This strongly suggests that carbonatites contain juvenile strontium derived from the deep crust or from the mantle. It is theoretically possible that carbonatites contain juvenile strontium, but recycled (once sedimentary) carbon dioxide. However, the work of ROEDDER (1965) on fluid inclusions suggests that the carbon dioxide in carbonatites is also juvenile and that it may well come from the mantle. The strontium isotopic data appear to form one more link in the chain of chemical and petrologic evidence showing that carbonatites are igneous rocks that come from great depth, probably from the Earth's mantle.

3. Alkalic Rocks

In Table VI.1 we list the $^{87}Sr/^{86}Sr$ ratios and the mean rubidium and strontium contents of a number of alkalic rock suites. Fig. VI.2, VI.3, and VI.4 are graphs of initial $^{87}Sr/^{86}Sr$ ratio versus Rb/Sr ratio, versus rubidium content, and versus strontium content, respectively, for the data shown in Table VI.1; they are modified from figures given by POWELL and BELL (1972). Also included for reference on the graphs are the fields for oceanic basaltic rocks and for continental basaltic rocks, based on the literature survey referred to in the caption to Fig. IV.1.

Table VI.1. $^{87}Sr/^{86}Sr$ and rubidium/strontium ratios, and rubidium and strontium contents of alkalic rocks

Locality: rock type	Number of samples	Initial $^{87}Sr/^{86}Sr$	Rb/Sr	Rb (ppm)	Sr (ppm)	Source of data
Africa						
(a) Eastern Rift						
carbonatite	6	0.7034	0	0	5306	1
ijolite	10	0.7045	0.10	52.5	527	1
nephelinite	9	0.7042	0.041	49.6	1211	1
fenite (not shown on graphs)	8	0.7058 — 0.7499				

Table VI.1 (continued)

Locality: rock type	Number of samples	Initial $^{87}Sr/^{86}Sr$	Rb/Sr	Rb(ppm)	Sr(ppm)	Source of data
(b) Western Rift						
absarokite	6	0.7076	0.141	138	1015	2
banakite	4	0.7074	0.134	175	1316	2
basanite	2	0.7073	0.104	127	1249	2
carbonated lava	4	0.7044	0	0	7240	2
katungite	5	0.7047	0.052	124	2449	2
kivite	8	0.7067	0.107	111	1047	2
latite	4	0.7099	0.339	207	613	2
leucitite	23	0.7058	0.102	152	1713	2
mafurite	6	0.7050	0.059	129	2256	2
melilitite	3	0.7057	0.098	163	1670	2
murambite	5	0.7070	0.118	105	886	2
nephelinite	17	0.7047	0.058	141	2472	2
shoshonite	4	0.7081	0.158	152	1048	2
ugandite	6	0.7056	0.093	135	1808	2
Australia						
West Kimberley	8	0.7169	0.287	402	1393	3
Europe						
Jumilla, Spain	6	0.7148	0.097	144	1524	3
Laacher See, Germany	9	0.7054	0.093	151	1625	7, 8
Oslo Province	28	0.7041	—	—	—	5
Roman Province	22 (18[a])	0.7092	0.49[a]	381	839	4
North America						
Absaroka Field, Wyoming	7	0.7059	0.159	132	1018	9
Bearpaw Mountains	7	0.7076	0.183	200	1196	3
Highwood Mountains	10	0.7078	0.148	159	1370	3
Hopi Buttes	6	0.7050	0.006	10.5	1889	3
Leucite Hills	17	0.7063	0.109	268	2625	3
Montana diatremes	13	0.7050	0.075	113	2005	3
Monteregian Hills	13	0.7040	0.037	—	—	6
Navajo Province	11	0.7069	0.103	127	1305	3

[a] Four samples from Ischia are omitted from calculation of the average Rb/Sr ratio because of their unusually high Rb/Sr ratios.

Sources of Data (see Bibliography)

1. BELL, K., and J. L. POWELL, 1970.
2. BELL, K., and J. L. POWELL, 1969.
3. POWELL, J. L., and K. BELL, 1970.
4. HURLEY, P. M., H. W. FAIRBAIRN, and W. H. PINSON, 1966.
5. HEIER, K. S., and W. COMPSTON, 1969 b.
6. FAIRBAIRN, H. W., G. FAURE, W. H. PINSON, P. M. HURLEY, and J. L. POWELL, 1963.
7. HOEFS, J. and K. H. WEDEPOHL, 1968.
8. Unpublished data of the authors.
9. PETERMAN, Z. E., B. R. DOE, and H. J. PROSTKA, 1970.

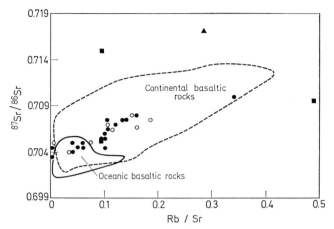

Fig. VI.2. Plot of initial ^{87}Sr/^{86}Sr ratio versus Rb/Sr ratio for most of the alkalic rock suites listed in Table VI.1. Also shown for comparison are fields for oceanic basaltic rocks and continental basaltic rocks. To maintain a reasonable scale a few continental basaltic rocks with Rb/Sr ratios exceeding 0.5 have been omitted. Symbols used are: ● = alkalic rocks from the Eastern and Western Rifts of Africa; ○ = alkalic rocks from North America; ■ = alkalic rocks from Europe; ▲ = alkalic rocks from West Kimberley, Australia. The African and North American suites define a roughly-linear, positive trend between initial ^{87}Sr/^{86}Sr ratio and Rb/Sr ratio (r = 0.74). Since processes of fractional crystallization do not change ^{87}Sr/^{86}Sr ratios, variations in initial ^{87}Sr/^{86}Sr ratio, much less a positive correlation between initial ^{87}Sr/^{86}Sr and Rb/Sr ratios, cannot be explained by fractional crystallization alone

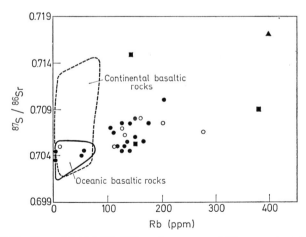

Fig. VI.3. Plot of initial ^{87}Sr/^{86}Sr ratio versus rubidium content for most of the alkalic rock suites listed in Table VI.1. See the caption of Fig. VI.2 for an explanation of the symbols used

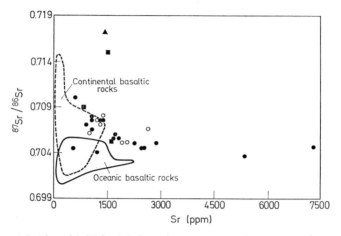

Fig. VI.4. Plot of initial $^{87}Sr/^{86}Sr$ ratio versus strontium content for most of the alkalic rock suites listed in Table VI.1 (see the caption for Fig. VI.2 for a listing of symbols used). The two points that fall within the field for oceanic basaltic rocks represent the ijolites and nephelinites from the Eastern Rift of Africa. The suites from North America, the Western Rift, and the Laacher See and Roman provinces appear to define a rough inverse relationship between the initial $^{87}Sr/^{86}Sr$ ratio and the strontium content

a) Potassic Rocks

A great deal of strontium isotope data are available for the potassic rocks of the Western Rift of Africa, and it will be instructive to consider them separately. BELL and POWELL (1969) discussed these data in more detail.

As shown in Table VI.1 and in Fig. VI.5, the initial $^{87}Sr/^{86}Sr$ ratios of the potassic rocks from the Western Rift are variable and correlate strongly with their Rb/Sr ratios. If the pattern shown in Fig. VI.5 is interpreted as an approximation to a Rb-Sr isochron, its slope corresponds to a date of about 500 million years. Since the rocks are only Pliocene to Recent in age, they must somehow have inherited a pattern of roughly-correlated $^{87}Sr/^{86}Sr$ and Rb/Sr ratios from an older rock system or systems. Of course all volcanic rocks inherit their constituent elements from their immediate parent materials, but ordinarily igneous processes produce isotopic homogenization and destroy any parent isotope-daughter isotope correlation that may have existed in the source material. Thus, as we pointed out in an earlier chapter, a cogenetic suite of igneous rocks usually crystallizes with variable Rb/Sr ratios but uniform $^{87}Sr/^{86}Sr$ ratios — a fact that is the basis of the whole-rock Rb-Sr dating method. Such isotopic homogenization has not happened to the rocks shown in Fig. VI.5; otherwise they would plot along a horizontal line.

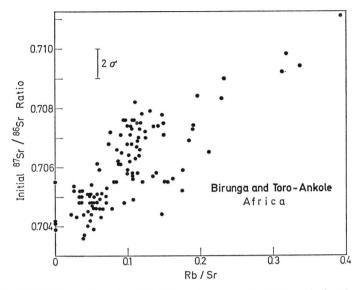

Fig. VI.5. Plot of the initial ^{87}Sr/^{86}Sr ratio versus the Rb/Sr ratio for the suites o alkalic rocks from the Western Rift of Africa. This diagram is modified from one by BELL and POWELL (1969). If the positive correlation shown is interpreted as an approximation to a Rb-Sr isochron, its slope corresponds to a date of about 500 million years. Yet rocks themselves are only Pliocene to Recent in age

BELL and POWELL (1969) thought that the correlation between ^{87}Sr/^{86}Sr and Rb/Sr ratios in the potassic rocks of the Western Rift could have resulted from (1) partial melting of old rocks, or (2) mixing of two appropriate materials, since such mixtures would plot along a straight line on a graph of ^{87}Sr/^{86}Sr against Rb/Sr. They attempted to test the mixing hypothesis by using the data shown in Fig. VI.2 to VI.5 and comparing the possible compositions of the hypothetical end members with those of actual rocks. For example, Fig. VI.5 shows that if only two components were involved, one (call it A) would have plotted at the extreme lower left; that is, it would have had low ^{87}Sr/^{86}Sr and Rb/Sr ratios, and the other (B) would have been relatively high in both and would have plotted at the upper right or off the scale of the diagram. Fig. VI.3 and VI.4 interpreted in a similar way show that A would have been low in rubidium and high in strontium and that B would have been high in rubidium and low in strontium. Many sialic rocks might have the properties required for B, but only strontium-rich ultrabasic rocks such as nephelinite and carbonatite have those required for A. BELL and POWELL therefore concluded that if the potassic rocks of the Western Rift were formed by mixing, one of the components was sialic and the other was a nephelinite or carbonatite magma. They pointed out that neither mixing nor partial melting provides a simple explanation of the high potassium contents of the rocks.

b) Sodic Rocks

Although they are more common than potassic rocks, fewer sodic rocks have been analyzed for $^{87}Sr/^{86}Sr$ ratio. Most of the rocks of carbonatite-alkalic rock complexes are richer in sodium than in potassium. The sodic provinces listed in Table VI.1 are the Eastern Rift of Africa; the Oslo Province, and the Hopi Buttes, the Montana diatremes, and the Monteregian Hills of North America. The $^{87}Sr/^{86}Sr$ ratios of these rocks are in the range 0.703 to 0.705 — lower than those of most of the potassic rocks listed.

POWELL et al. (1966) analyzed carbonatites and associated alkalic rocks from several complexes, found no significant differences in their $^{87}Sr/^{86}Sr$ ratios, and concluded that all of the rocks of an individual complex were comagmatic. However, their precision of measurement (± 0.002) would not have allowed them to detect very small differences, and they analyzed no fenites.

BARKER and LONG (1969) used geochemical and strontium isotope data to show that a nepheline and analcime-bearing syenite from Brookville, New Jersey, was produced by mixing of granophyric liquid and argillite.

MITCHELL and CROCKET (1969) analyzed several specimens from the classic carbonatite-alkalic rock complex at Fen, Norway. They found that the rocks do not yield a Rb-Sr isochron. When the present-day $^{87}Sr/^{86}Sr$ ratios were corrected according to the known K-Ar age of the complex, the rocks were found to have different initial $^{87}Sr/^{86}Sr$ ratios, with the carbonatite having the lowest. MITCHELL and CROCKET concluded that it was the primary magma and that the silicate rocks had been produced by fenitization.

BELL and POWELL (1970) analyzed a number of specimens from the alkalic complexes of the Eastern Rift in Uganda. A summary of their results is included in Table VI.1. They also found that the carbonatites had lower ratios than the alkalic silicate rocks. No patterns characteristic of mixing were found, and even the phonolites and trachytes had ratios lower than those expected of remobilized fenites. They found that the rocks actually classified as fenites on the basis of field and petrographical evidence had very high initial $^{87}Sr/^{86}Sr$ ratios — reaching 0.750 in one case. This corroborated the conclusion of field geologists that these were old country rocks that had been metasomatized.

c) Comparison of North American and African Alkalic Rocks

If the alkalic rocks from Europe and Australia shown on Fig. VI.3 are ignored, those that are left (from Africa and North America) form a rather tightly-clustered group with an apparent positive correlation between Rb/Sr and $^{87}Sr/^{86}Sr$ ratios. In fact, the correlation coefficient (r) between those parameters for the African and North American alkalic rocks is 0.74, a value

significant at greater than 99 percent confidence limits. This similar co-
variance of $^{87}Sr/^{86}Sr$ and Rb/Sr ratios in alkalic rocks from two continents
suggests that they all have been formed by the same process. Since it seems
extremely unlikely that any mixing processes at such widely separated
localities would produce similar Rb/Sr and $^{87}Sr/^{86}Sr$ ratios, we believe that
partial melting of an older, deeper layer — probably the upper mantle — may
have produced these alkalic rocks. PETERMAN and HEDGE (1971) (see Chapter
IV) argued that the more potassic and ^{87}Sr-rich lavas from oceanic islands had
been derived from the more alkali-rich (and most primitive) zones of the
mantle. Their argument can be extended to the rocks being considered here:
the potassic rocks from the Western Rift, with their initial $^{87}Sr/^{86}Sr$ ratios of
up to 0.711, could have been derived from correspondingly alkali-rich,
perhaps phlogopite-bearing, and least-depleted parts of the upper mantle.
Effects of contamination, superimposed on these primary alkali and ^{87}Sr-
abundances, could explain the very high initial $^{87}Sr/^{86}Sr$ ratios of the potassic rocks from Jumilla, Spain, and West Kimberley, Australia.

4. Summary

Strontium isotopic studies of carbonatites have added further evidence
to the view that these rocks are magmatic and derived from great depth,
possibly from the Earth's mantle. The initial $^{87}Sr/^{86}Sr$ ratios of carbonatites
are low and similar to those of oceanic basalts; limestones of Paleozoic and
younger age, by contrast, have higher ratios.

The petrogenesis of the alkalic rocks remains an important but unsolved
problem. Strontium isotopic data have shown that many of these rocks are
formed neither by limestone syntexis nor by simple fractional crystallization of normal basaltic or granitic magmas. Some strontium isotope evidence
suggests that the potassic rocks could have formed by a mixing process, but
the data can perhaps be explained equally well by the hypothesis of partial
melting of a very primitive mantle.

VII. Ultramafic and Related Rocks

1. Introduction

The ultramafic rocks are composed mainly of dark-colored mafic or ferro-magnesian minerals such as olivine, pyroxene, amphibole, serpentine, garnet, biotite, and opaque oxides. They are usually, but not always, ultra-basic (having less than 45 percent SiO_2). Ultramafic rocks have been compre-hensively reviewed in a volume edited by WYLLIE (1967), who recognizes 10 ultramafic rock associations. $^{87}Sr/^{86}Sr$ ratio measurements are available for the following: (1) the layered gabbro-norite-peridotite association in major intrusions, (2) the alpine-type peridotite-serpentinite association, (3) ultramafic rocks in differentiated basic sills and in minor intrusions, (4) kimberlites, (5) ultramafic nodules, (6) alkalic ultrabasic rocks in ring complexes, and (7) ultrabasic lavas. In the following sections of this chapter we shall review the strontium isotopic data for the first five of these associa-tions; the last two have been covered in Chapter VI.

Ultramafic rocks, especially those occurring as nodules in basalts and in kimberlites, have received a great deal of attention, and indeed may be of much greater importance than their scarcity would suggest. Among all of the rocks known, only the ultramafic rocks dunite, eclogite, and peridotite have geophysical properties that match those of the upper mantle. Most geologists have concluded from this and other evidence that the mantle itself is made up of ultramafic rock. Thus the peridotite nodules in oceanic basalts, for example, conceivably could be actual pieces of the upper mantle.

2. $^{87}Sr/^{86}Sr$ Ratios of Ultramafic Rocks

a) The Layered Gabbro-Norite-Peridotite Association in Major Intrusions

In several different areas of the world there occur very large, differ-entiated gabbroic lopoliths or stratiform intrusions, perhaps best ex-emplified by the Bushveld Complex of South Africa and the Stillwater Com-plex of Montana. Somewhat smaller funnel-shaped ultramafic bodies, like the Skaergaard Complex of East Greenland, also belong to this association. Some of these rocks are very old, and because of their low Rb/Sr ratios, the accurate determination of their age has been a difficult analytical problem.

The available strontium isotope data are shown in Table VII.1. The initial $^{87}Sr/^{86}Sr$ ratios vary from about 0.703 to 0.712, but most are in-distinguishable from the ratios of the majority of basalts. This evidence is

Table VII.1. Initial $^{87}Sr/^{86}Sr$ ratios of rocks of the layered gabbro-norite-peridotite association in major intrusions

Locality	Initial $^{87}Sr/^{86}Sr$	Ref.
Bushveld Complex	0.7065	DAVIES et al. (1970)
Duluth gabbro	0.7055	FAURE et al. (1969)
Endion Sill	0.7046	FAURE et al. (1969)
Great Dyke	0.7024	DAVIES et al. (1970)
Losberg Complex	0.7064	DAVIES et al. (1970)
Nipissing diabase	0.7060	VAN SCHMUS (1965), FAIRBAIRN et al. (1969)
Northeast Scotland	0.7032 — 0.7122	PANKHURST (1969)
Skaergaard Complex		HAMILTON (1964)
Marginal Border Group	0.7063[a]	
Layered Series	0.7066[a]	
Melanogranophyres	0.7071[a]	
Stillwater Complex	0.7029	FENTON and FAURE (1969)
Sudbury Irruptive	0.704 —0.705	FAURE et al. (1964)
Trompsburg Complex	0.7043	DAVIES et al. (1970)
Usushwana Complex	0.7031	DAVIES et al. (1970)

[a] Probably require downward correction to express relative to $^{87}Sr/^{86}Sr$ = 0.7080 for the Eimer and Amend interlaboratory standard.

consistent with the generally-accepted view that large mafic and ultramafic intrusions crystallize from basaltic magmas.

The mafic and ultramafic rocks of Northeast Scotland are exceptional among those listed in Table VII.1, in having $^{87}Sr/^{86}Sr$ ratios as high as 0.712. PANKHURST (1969) found that in one of the gabbroic intrusions, the Insch Mass, ferrogabbros and syenogabbros consistently had initial $^{87}Sr/^{86}Sr$ ratios in the range 0.711 to 0.712, although the ratios of the other rocks of the mass were as low as 0.703. Furthermore, the initial $^{87}Sr/^{86}Sr$ ratios of the Insch rocks were found to increase with decreasing anorthite content of their plagioclases. Therefore, the $^{87}Sr/^{86}Sr$ ratio of the parent magma seems to have changed as crystallization proceeded.

Based on the large amount of age and chemical information available for the gabbros and their surrounding country-rock, PANKHURST (1969) was able to conclude that the relatively high $^{87}Sr/^{86}Sr$ ratios of the gabbros were not due to bulk assimilation of country rock, to wall-rock reaction (GREEN and RINGWOOD, 1967), or to selective extraction of ^{87}Sr from country-rock micas. He considered the only possible mechanism to be one of isotope exchange or equilibration between basic magma and country rock that would have progressively raised the $^{87}Sr/^{86}Sr$ ratio of the magma and lowered the $^{87}Sr/^{86}Sr$ ratio of the country rock. Isotope equilibration would probably be most effective if water were present to act as a medium for the exchange, and it could be important in determining the $^{87}Sr/^{86}Sr$ ratios of other, less strontium-rich, serpentinized ultramafic rocks.

These results show that high initial $^{87}Sr/^{86}Sr$ ratios can occur in mafic rocks that appear to be of deep origin. Therefore, a high initial ratio in a rock cannot entirely eliminate the possibility that its parent magma was derived from the basal crust or from the upper mantle (see also Chapter VI). Furthermore, the presence of variations among the initial $^{87}Sr/^{86}Sr$ ratios of a suite of apparently cogenetic rocks in some cases could reflect isotope equilibration, and not some more fundamental magmatic or assimilative process.

Some major stratiform intrusions and some basic sills contain an uppermost layer or zone of acid granophyre whose origin often has been ascribed to fractional crystallization. However, HAMILTON (1963) found that three specimens of acid granophyre from the Skaergaard Complex had initial $^{87}Sr/^{86}Sr$ ratios of 0.7094, 0.7141, and 0.7303 — substantially higher than the ratios of the mafic rocks of the Complex. His data also show that the strontium content and the initial $^{87}Sr/^{86}Sr$ ratio are inversely related in the three specimens, supporting his conclusion that the Skaergaard granophyre was contaminated. Further evidence for this conclusion comes from the work of TAYLOR and EPSTEIN (1963) with oxygen isotopes.

DAVIES et al. (1970) analyzed associated acid and mafic rocks from several layered mafic intrusions from Southern Africa. In each the $^{87}Sr/^{86}Sr$ ratios of the acid and mafic types overlapped, although only slightly so for the Bushveld Complex.

Anorthosites: These rocks are composed mainly of calcic plagioclase with minor pyroxene, or olivine, or both. They sometimes occur as distinct layers within stratiform ultramafic intrusions, and although they are not ultramafic, their geologic occurrence makes it logical to consider them as related to ultramafic rocks.

The other and more important geologic occurrence of anorthosites is as large (sometimes batholithic) independent intrusions or massifs in Precambrian terranes, frequently in close association with pyroxene-bearing rocks such as charnockite and pyroxene syenite. HERZ (1969) has shown that when the continents are arranged in a pre-drift reconstruction, these anorthosite massifs lie in two long belts, one in Laurasia and the other in Gondwanaland. HERZ also noted that ages of almost all anorthosite massifs that have been dated fall in the range 1300 ± 200 millon years.

TURNER and VERHOOGEN (1960) and HEATH (1967) reviewed the petrogenesis of anorthosites. Some writers had proposed that anorthosites form through metamorphism, metasomatism, or anatexis of argillaceous limestones or other sedimentary rocks. If so, their initial $^{87}Sr/^{86}Sr$ ratios would be expected to be relatively high (see Chapter VIII), unless the metasomatizing solutions had very high strontium contents and low isotopic ratios. HEATH (1967) analyzed 57 individual anorthosite specimens from 15 different areas, including the Adirondack Mountains; Laramie, Wyoming; Nain, Labrador; Morin, Quebec; Naero Fjord, Norway, and others. The initial ratios of each of the 15 bodies fell between 0.703 and 0.706. Such a narrow range and low values make it extremely unlikely that the anorthosites analyzed formed by any process involving sedimentary material. It is consistent with anorthosites forming by crystal settling from basaltic magmas, or by crystallization of a melt of anorthositic composition derived from the lower crust or from the mantle. HEATH's work did not resolve the question of whether the anorthosites and associated pyroxenic rocks such as charnockite and pyroxene syenite are genetically related. He found Adirondack anorthosite and pyroxene syenite to have identical initial ratios, but not enough samples of each rock type from other areas have been analyzed to provide an answer to this problem.

b) The Alpine-Type Peridotite-Serpentinite Association

The folded, dissected mountain belts of the world typically contain many small lenses or sheets of dunite, peridotite, and serpentinite arranged in elongate chains. For example, hundreds of such bodies run down the axis of the Appalachian Mountains from the Gaspé Peninsula to North Carolina (see TURNER and VERHOOGEN, pp. 307—321). The controversy surrounding the origin of these rocks is one of the best examples in the whole field of geology of the discrepancy between laboratory data (which showed that ultrabasic magmas could only exist at very high temperatures) and field observation (which showed that the ultramafic rocks, though intrusive, had been emplaced at low temperatures). WYLLIE (1967, pp. 407—413) has given a summary of this argument and of recent field and laboratory work bearing on it. With the advent of plate tectonic theory, more and more authors seem to be regarding the alpine-type ultramafic rocks as pieces of sea-floor squeezed-up without complete melting during plate collision. The strontium isotope data that are available for dunites, peridotites, and serpentinites are listed in Table VII.2.

Studies of the initial $^{87}Sr/^{86}Sr$ ratios of alpine-type ultramafic rocks were made by ROE (1964), who found values from 0.705 to 0.715, and by STUEBER and MURTHY (1966), who observed a range from 0.706 to 0.729 (see HURLEY, 1967). Both authors rejected the possibility that the rocks had been contaminated, and agreed that the high initial $^{87}Sr/^{86}Sr$ ratios of many indicate that they come from a zone that separated from the rest of the mantle a billion or more years ago.

Although for the most part ROE, and STUEBER and MURTHY, analyzed different specimens, their data for the rubidium and strontium concentrations and the $^{87}Sr/^{86}Sr$ ratios of ultramafic rocks agreed closely, with one

Table VII.2. $^{87}Sr/^{86}Sr$ and rubidium/strontium ratios, and rubidium and strontium contents of alpine-type dunites, peridotites, and serpentinites

Rock type	$^{87}Sr/^{86}Sr$	Rb(ppm)	Sr(ppm)	Rb/Sr	Locality	Reference
Oceanic						
Peridotite	0.7066	8.35	13.7	0.609	New Caledonia	Roe (1964)
Dunite	0.7079	0.073	0.523	0.15	New Caledonia	Roe (1964)
Dunite	0.7127	0.147	0.499	0.295	New Caledonia	Roe (1964)
Dunite	0.7078	0.302	6.28	0.048	Papua	Stueber and Murthy (1966)
Serpentinite	0.7053	0.048	6.85	0.007	Puerto Rico	Roe (1964)
Serpentinite	0.7058	0.041	4.92	0.0083	Puerto Rico	Roe (1964)
Serpentinite	0.7067	0.298	7.71	0.0386	Puerto Rico	Roe (1964)
Serpentinite	0.7086	0.09	1.3	0.069	Puerto Rico	Hart (1964)
Serpentinite	0.7079	0.27	2.8	0.096	Puerto Rico	Hart (1964)
Peridotite	0.7069	0.55	11.5	0.048	Equatorial Mid-Atlantic Ridge	Bonatti et al. (1970)
Peridotite	0.7063	0.50	16.9	0.030	Equatorial Mid-Atlantic Ridge	Bonatti et al. (1970)
Peridotite	0.7227	0.49	9.6	0.051	Equatorial Mid-Atlantic Ridge	Bonatti et al. (1970)
Peridotite	0.7096	0.22	64.7	0.003	Equatorial Mid-Atlantic Ridge	Bonatti et al. (1970)
Peridotite	0.7114	<0.1	6.3	<0.016	Equatorial Mid-Atlantic Ridge	Bonatti et al. (1970)
Peridotite	0.7076	0.77	7.6	0.101	Equatorial Mid-Atlantic Ridge	Bonatti et al. (1970)
Peridotite	0.7111	2.28	18.5	0.123	Equatorial Mid-Atlantic Ridge	Bonatti et al. (1970)
Peridotite	0.7089	0.48	7.5	0.064	Equatorial Mid-Atlantic Ridge	Bonatti et al. (1970)
Peridotite	0.7138	0.10	4.1	0.024	Equatorial Mid-Atlantic Ridge	Bonatti et al. (1970)
Peridotite	0.7113	—	6.4	—	Equatorial Mid-Atlantic Ridge	Bonatti et al. (1970)
Continental						
Peridotite	0.7043	—	—	—	Alaska, Duke Island	Lanphere (1968)
Dunite	0.7068	—	—	—	Alaska, Blashke Island	Lanphere (1968)
Dunite	0.7101	0.072	2.32	0.031	Alaska	Stueber and Murthy (1966)
Dunite	0.7030	1.80	1.50	1.20	Alaska, Duke Island	Stueber and Murthy (1966)

Table VII.2. (continued)

Rock type	$^{87}Sr/^{86}Sr$	Rb(ppm)	Sr(ppm)	Rb/Sr	Locality	Reference
Serpentinite	0.7152	0.284	6.89	0.041	Australia	STUEBER and MURTHY (1966)
Peridotite	0.7036	0.099	21.1	0.005	Australia, Camperdown	STUEBER (1969)
Dunite	0.7100	0.130	10.2	0.013	British Columbia	STUEBER and MURTHY (1966)
Dunite	0.7290	2.42	14.7	0.165	Greenland	STUEBER and MURTHY (1966)
Dunite	0.7063	0.099	8.10	0.012	Japan	STUEBER and MURTHY (1966)
Serpentinite	0.7109	1.04	8.36	0.124	Newfoundland	STUEBER and MURTHY (1966)
Dunite	0.7091	0.111	4.39	0.025	New Zealand	STUEBER and MURTHY (1966)
Dunite	0.7156	0.077	2.99	0.026	North Carolina	STUEBER and MURTHY (1966)
Peridotite	0.7109	0.04	2.1	0.019	North Carolina, Addie	STUEBER (1967)
Dunite	0.7175	0.034	0.406	0.084	North Carolina, Balsam Gap	STUEBER (1969)
Dunite	0.7165	0.017	1.65	0.010	North Carolina, Balsam Gap	STUEBER (1969)
Serpentinite	0.7177	0.054	1.25	0.043	North Carolina, Balsam Gap	STUEBER (1969)
Dunite	0.7125	0.079	0.803	0.097	North Carolina, Balsam Gap	ROE (1964)
Dunite	0.7067	0.084	5.17	0.016	North Carolina, Buck Creek	STUEBER (1969)
Peridotite	0.7241	0.227	1.0	0.206	North Carolina, Dark Ridge	STUEBER (1967)
Dunite	0.7275	0.037	1.60	0.023	North Carolina, Dark Ridge	STUEBER (1969)
Dunite	0.7136	0.021	0.321	0.065	North Carolina, Addie	STUEBER (1969)
Dunite	0.7126	0.077	2.99	0.026	North Carolina, Addie	STUEBER and MURTHY (1966)
Dunite	0.7078	0.131	9.89	0.013	Norway	STUEBER and MURTHY (1966)
Peridotite	0.7105	0.158	5.52	0.029	Quebec	STUEBER and MURTHY (1966)
Peridotite	0.7149	0.380	0.787	0.484	Quebec, Black Lake	ROE (1964)
Peridotite	0.7096	0.307	3.60	0.089	Quebec, Black Lake	ROE (1964)
Dunite	0.7087	0.094	4.84	0.019	Russia, Urals	ROE (1964)
Serpentinite	0.7077	0.083	0.914	0.091	Russia, Urals	ROE (1964)
Serpentinite	0.7087	0.146	3.33	0.043	Russia, Urals	ROE (1964)
Dunite	0.7272	0.140	4.28	0.033	Turkey	STUEBER and MURTHY (1966)
Peridotite	0.7084	0.093	3.89	0.024	Venezuela	STUEBER and MURTHY (1966)
Serpentinite	0.7104	0.201	1.46	0.138	Vermont, Lowell	ROE (1964)

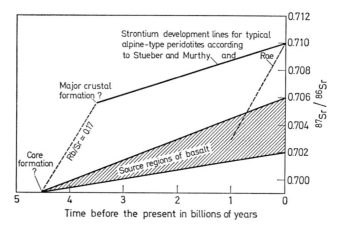

Fig. VII.1. The $^{87}Sr/^{86}Sr$ ratio development diagram for alpine-type ultramafic rocks according to the model of ROE (1964) and the model of STUEBER and MURTHY (1966). The diagram is modified from those shown by STUEBER and MURTHY (1966) and HURLEY (1967). Also included is a shaded zone that shows the growth of $^{87}Sr/^{86}Sr$ ratio in the source regions of oceanic basalts with time (cf. Figs. III.1, V.1). According to both models the alpine-type ultramafic rocks have been derived from a residual zone in the mantle with little if any accompanying change in their Rb/Sr and $^{87}Sr/^{86}Sr$ ratios. The very low strontium contents of alpine-type ultramafic rocks, however, render them extremely susceptible to contamination with radiogenic strontium, and if they have been contaminated, their present Rb/Sr and $^{87}Sr/^{86}Sr$ ratios are not directly related to those of their deep source regions. It may be unwise to use the initial $^{87}Sr/^{86}Sr$ ratios of alpine-type ultramafic rocks as the basis for speculation about the geochemical evolution of the Earth

exception: ROE obtained strontium contents for dunites that were about 10 times lower than those measured by STUEBER and MURTHY. This caused the Rb/Sr ratios of dunites as reported by ROE to be about 10 times higher than those given by STUEBER and MURTHY, and this in turn led them to radically different conclusions regarding the history of ultramafic material in the mantle. Their interpretations are illustrated in Fig. VII.1, where $^{87}Sr/^{86}Sr$ development lines are plotted for the mantle (after Fig. III.1), for ultramafic rocks according to ROE (1964), and for ultramafic rocks according to STUEBER and MURTHY (1966). HURLEY (1967) has presented a similar diagram.

The lower Rb/Sr ratios obtained by STUEBER and MURTHY (1966) caused their strontium development line for alpine-type ultramafic rocks to have a correspondingly low slope [see Eq. (II.16)] and to approximately parallel the strontium development line for the mantle. They proposed that very early in the Earth's history an ultramafic layer formed with a relatively high Rb/Sr ratio (they suggest 0.17) that caused it to develop significant amounts of radiogenic strontium. With the separation of the crust about 3500 million years ago and the relative concentration of rubidium in it, the Rb/Sr ratio of this residual ultramafic layer was lowered to the value now observed in alpine-type ultramafic rocks. Thus, according to their model, the high $^{87}Sr/^{86}Sr$ ratios of the ultramafic rocks were almost completely attained during the first 1000 million years of the Earth's history.

The higher Rb/Sr ratios obtained by ROE (1964) caused his ultramafic-rock development line to intersect that for the mantle at a point corresponding to an age of about one billion years (see HURLEY, 1967). ROE's interpretation was therefore that the parent material of alpine-type ultramafic rocks separated from the mantle around one billion years ago, and that the high $^{87}Sr/^{86}Sr$ ratios of the rocks were generated as a result of their higher Rb/Sr ratios during this one-billion-year period. Thus both the model of STUEBER and MURTHY (1966) and that of ROE (1964) require that many ultramafic rocks have had a two-stage history and that they are derived from a zone which became segregated from the rest of the mantle at least a billion years ago. Recently BONATTI et al. (1970) came to a similar conclusion on the basis of high $^{87}Sr/^{86}Sr$ ratios (up to 0.723, as shown in Table VII.2) found in peridotites dredged from the Equatorial Mid-Atlantic ridge. BONATTI (1971) developed a detailed model for the evolution of the Atlantic Ocean and its suboceanic mantle based in large part on the observation that both continental and dredged oceanic peridotites can have unusually high $^{87}Sr/^{86}Sr$ ratios.

We believe that before the high initial $^{87}Sr/^{86}Sr$ ratios of ultramafic rocks can be used as a sound basis for speculation about the history of the Atlantic Ocean and of the whole Earth, it must be firmly established that they are not due to contamination. STUEBER and MURTHY (1966) wrote that there is no petrographic evidence for such contamination in the specimens they studied, and regarded it unlikely that contamination in diverse environments would produce the uniformly low Rb/Sr ratios observed in ultramafic rocks. BONATTI et al. argued against contamination in their dredged peridotites because (1) the $^{87}Sr/^{86}Sr$ ratios were higher than those of seawater, the most likely contaminant, (2) there was no correlation with strontium content, and (3) some samples that must have been exposed to seawater had normal ratios. However, the work of DASCH (1969a) (see Chapter VIII) showed that contaminants with $^{87}Sr/^{86}Sr$ ratios higher than those measured in any ultramafic rock (ocean-floor sediments) are common in the ocean basins.

With the exception of one sample with about 65 ppm, none of the rocks listed in Table VII.2 has more than about 21 ppm of strontium, and a number have less than 5 ppm. Fig. IV.1 shows how very low in strontium content are all of the dunites and peridotites with high $^{87}Sr/^{86}Sr$ ratios. Many of these ultramafic rocks are partly serpentinized and occur in orogenic belts where radiogenic strontium may have been available. It is difficult to believe that even partial serpentinization of a strontium-poor ultramafic rock in a water-rich orogenic belt or on the sea floor could fail to affect its $^{87}Sr/^{86}Sr$ ratio, and FENTON and FAURE (1969) found that two serpentinized specimens from the Stillwater Complex lay above the isochron formed by the non-serpentinized specimens. In any case, the process of isotope equilibration defined by PANKHURST (1968) would be especially effective in water-rich environments, and he found peridotites to have higher ratios than some of the gabbros.

STUEBER (1969) in a study of several ultramafic complexes in North Carolina found that the $^{87}Sr/^{86}Sr$ ratios of five closely-associated alpine-type intrusions spanned the entire range previously reported by ROE (1964) and STUEBER and MURTHY (1966). The ratios also cover the range for dredged peridotites reported by BONATTI et al. (1970), though neither they nor BONATTI (1971) cite STUEBER's 1969 paper as a reference. STUEBER's results (1969) suggest that the variation in isotopic composition found in these ultramafic rocks is not a primary feature. Additional evidence comes from study of the chemical and isotopic compositions of the individual minerals and groundmass of ultramafic rocks:

GRIFFIN and MURTHY (1968, 1969) found a material imbalance between the amounts of potassium, rubidium, barium, and strontium in whole-rock eclogites and garnet peridotites and in their constituent minerals, and concluded that substantial fractions of these four elements were present in secondary minerals or in alteration products. COMPSTON and LOVERING (1969) found that a material balance existed for strontium within eclogite and garnet peridotite inclusions, but not for the alkali elements. However, they also found that the major minerals within a single inclusion had not been in isotopic equilibrium. ALLSOPP et al. (1969) discovered that the intergranular material in eclogitic and peridotitic inclusions in kimberlite had both a higher $^{87}Sr/^{86}Sr$ ratio and a higher strontium content than the primary minerals, and they suggested that the intergranular strontium had entered the inclusions during the emplacement of the pipes.

These studies all show that part of the strontium in some ultramafic rocks is in secondary minerals. There appear to be no petrographic or chemical criteria, save for analysis of the $^{87}Sr/^{86}Sr$ ratios and strontium contents of whole-rocks and all of their minerals, by which ultramafic rocks that contain only primary strontium, and no secondary strontium, can be recognized. We agree with STUEBER (1969): "... it seems that the present Rb-Sr systems of these alpine-type ultramafics have no direct relationship with the mantle." We would extend his conclusion to the peridotites analyzed by BONATTI et al. (1970). At least it seems fair to state that the burden of proof that such a relationship exists should rest with those who draw large-scale geological conclusions from the rocks.

c) Ultramafic Rocks in Differentiated Basic Sills and in Minor Intrusions

Only a few representatives of this association have been analyzed for initial $^{87}Sr/^{86}Sr$ ratio. FAURE and HURLEY (1963) listed analyses for the Palisades Sill of New Jersey, and MOORBATH and BELL (1965) gave results for several mafic and ultramafic intrusive masses on Skye. All of these rocks were found to have initial ratios similar to those of most basalts.

d) Kimberlites

Kimberlite is an ultrabasic, brecciated rock consisting of angular grains of olivine (often serpentinized), and less abundant pyroxene, garnet, and mica set in a groundmass of serpentine and calcite. A variety of accessory minerals, sometimes including diamond, occur. Many authors believe that much of the CO_2 in kimberlites is primary, and there is considerable evidence that the diatremes now occupied by kimberlites were drilled out by a gas-rich magma.

Many writers believe that kimberlites are derived from very great depths in the mantle. Their chemistry is reminiscent of that of the potassic ultra-basic lavas covered in Chapter VI: both are unusually rich for ultramafic rocks in alkalis, Ti, P, Ba, Sr, Zr, etc. (DAWSON, 1967). This caused DAWSON (following HOLMES's earlier suggestions for the ultrabasic potassic lavas) to propose that kimberlites form by reaction between carbonatitic fluids and older sialic material.

It is apparent that different $^{87}Sr/^{86}Sr$ ratios would be expected to occur in kimberlites if they are entirely derived from deep in the mantle than if they form from a mixture of carbonatite and sial. However, application of stron-tium isotopic studies to kimberlite petrogenesis is not at all straightforward. This is because kimberlites contain a wide variety of foreign and cognate inclusions ranging in size from xenocrysts to boulders. The presence of a few old mica xenocrysts, for example, might raise the $^{87}Sr/^{86}Sr$ ratio of a kimberlite significantly. Although all authors who have analyzed kimberlites have attempted to avoid such inclusions and xenocrysts, since they range down to microscopic size it is impossible to state categorically that none at all were present in a given sample.

MITCHELL and CROCKET (1971) have made the most thorough study of the initial $^{87}Sr/^{86}Sr$ ratios of kimberlites; they observed a range from 0.705 to 0.718. They concluded that kimberlite may be formed by partial melting of garnet mica peridotite and that the higher ratios may have resulted from the occurrence of a larger proportion of radiogenic strontium from mantle phlogopite in some kimberlites.

e) Ultramafic Nodules

Peridotites and other ultramafic rocks are sometimes found as inclusions in volcanic rocks. Such inclusions could be cumulates of early-formed

Table VII.3. $^{87}Sr/^{86}Sr$ ratios of some peridotite inclusions and basaltic host rocks

Inclusion	Basaltic host rock[a]	Locality	Reference
0.7060	0.7035	Massif Central, France	LEGGO and HUTCHISON (1968)
0.7093	0.7038	Massif Central	LEGGO and HUTCHISON (1968)
0.7106	0.7036	Massif Central	LEGGO and HUTCHISON (1968)
0.7061	0.7043	Massif Central	LEGGO and HUTCHISON (1968)
0.7036	0.7035	Galapagos Islands	STUEBER and MURTHY (1966)
0.7064	0.7054	Hawaiian Islands	STUEBER and MURTHY (1966)
0.7036	0.7039	Australia	STUEBER (1969)
0.7057		Massif Central	LEGGO and HUTCHISON (1968)
0.7096		Massif Central	LEGGO and HUTCHISON (1968)
0.7090		Massif Central	LEGGO and HUTCHISON (1968)
0.7059		Massif Central	LEGGO and HUTCHISON (1968)
0.7037		Massif Central	LEGGO and HUTCHISON (1968)
0.7080		Kerguelen Islands	STUEBER and MURTHY (1966)
0.7058		Antarctica	STUEBER and MURTHY (1966)
0.7067		Austria	STUEBER and MURTHY (1966)
0.7045		Mexico	STUEBER and MURTHY (1966)
0.7062		California	STUEBER and MURTHY (1966)
0.7046		Tanzania	STUEBER and MURTHY (1966)
0.7064		South Africa	STUEBER and MURTHY (1966)
0.7049		South Africa	STUEBER and MURTHY (1966)
0.7083		New Zealand	STUEBER and MURTHY (1966)

[a] No analysis of the host rock was reported in most cases.

minerals from basalt magma, or they could be pieces of the upper mantle that were included in, but were genetically unrelated to, the volcanic rocks that enclose them. In the first case, inclusion and host basalt should have the same initial isotopic composition of strontium. In the second, their $^{87}Sr/^{86}Sr$ ratios might differ. The available strontium isotopic data for peridotite inclusions, and where available for their host basalts, are shown in Table VII.3.

The $^{87}Sr/^{86}Sr$ ratios of the peridotites range from 0.7036 to 0.7106; both of the extreme values occur in samples from the Massif Central of France. The initial $^{87}Sr/^{86}Sr$ ratios of all of the Massif Central basalts analyzed by LEGGO and HUTCHISON (1968), which include some not listed in Table VII.3, range from 0.7024 to 0.7049. Since the ultramafic inclusions from the Massif Central have significantly higher $^{87}Sr/^{86}Sr$ ratios than their host basalts, LEGGO and HUTCHISON concluded that the two types are not genetically related but were ". . . brought together by chance." They did point out, however, that the $^{87}Sr/^{86}Sr$ ratios of the inclusions could have been raised by contamination with radiogenic strontium, the problem we identified in section b of this chapter.

STUEBER and MURTHY (1966) studied two peridotite inclusion-host basalt pairs, and found that inclusion and host basalt had similar strontium isotopic compositions in both cases. O'NEIL et al. (1970) made the same observation for pyroxenite xenoliths enclosed in melilite basalt on Oahu. Agreement between the initial $^{87}Sr/^{86}Sr$ ratios of inclusion and host rock permits but does not prove the hypothesis that the two are cogenetic. As STUEBER and MURTHY pointed out, even if such a genetic relationship were indicated, it would still not be possible from the isotopic evidence to decide whether the inclusions were cumulates from the basalt magmas, pieces of the mantle that give rise to the basalts by partial melting, or the refractory residues of the basalts.

3. Summary

In reviewing the strontium isotopic data for ultramafic rocks, two different categories can be distinguished:

(1) Ultramafic and associated mafic rocks relatively high in strontium content. Examples are the rocks of the large stratiform intrusions, the anorthosites, and mafic and ultramafic rocks occurring in smaller differentiated sills. All of these rocks usually have low initial $^{87}Sr/^{86}Sr$ ratios identical to those of basalts. This result is consistent with the view that the parent magmas of such rocks were derived from the source regions of basalts. The gabbros of Northeast Scotland and the kimberlites are exceptions to this rule, but plausible explanations of the higher ratios of these two types have been developed.

(2) Ultramafic rocks with low strontium contents, such as the alpine-type peridotites and some peridotite nodules in basalts. These also may have low $^{87}Sr/^{86}Sr$ ratios, but many have ratios over 0.710, and they range as high as 0.729. The simplest explanation, though not necessarily the correct one, is that these strontium-poor rocks were contaminated with radiogenic strontium and perhaps with rubidium.

Strontium isotope analyses of rocks of both categories can yield useful petrogenetic information, providing that independent evidence can be presented to show whether or not the strontium in representatives of the second category is primary. The most useful kind of evidence would be the chemical and isotopic compositions of both the whole rocks and their individual mineral phases and groundmass.

VIII. Sedimentary Rocks and the Oceans

1. Introduction

Interest in the isotopic composition of strontium in sedimentary rocks is increasing rapidly. Recent work includes dating of argillaceous sedimentary rocks as well as studies of the variation of the $^{87}Sr/^{86}Sr$ ratios of the oceans in the geologic past. While much progress has been made, many challenging problems remain to be solved. In this chapter we shall examine what has already been done and point out where more progress is still to be made.

For the purpose of this discussion we shall distinguish between detrital and chemically-precipitated sedimentary rocks. Detrital sedimentary rocks are predominantly composed of mineral and rock particles derived by weathering of pre-existing rocks. Chemically precipitated rocks, on the other hand, consist primarily of authigenic minerals which form in the environment of deposition and whose chemical and isotopic composition may be controlled by the environment of deposition.

2. Marine Carbonate Rocks and the Oceans

The geochemistry of strontium in the oceans has been studied by ODUM (1957), LOWENSTAM (1964), TUREKIAN (1964), and others. The bulk of strontium entering the oceans is derived from marine limestones undergoing diagenesis and chemical weathering on the continents. Its average concentration in the oceans is about 8 milligrams per liter, making it one of the most abundant trace elements. The residence time of strontium in the oceans is 1.9×10^7 years, according to GOLDBERG (1965), which is long compared to the mixing rate of the oceans (approximately 10^3 years). Strontium is removed from the oceans primarily by co-precipitation of Sr^{2+} ions with calcium carbonate. The concentration of strontium in calcium carbonate depends on the mineralogy of the precipitating phase, the molar ratio of Sr^{2+} to Ca^{2+} in the water, the temperature, and a variety of other factors such as the composition and concentration of dissolved salts and biological effects. The molar ratio of Sr^{2+} to Ca^{2+} in the solid phase is related to that of the liquid phase by the distribution coefficient k:

$$k = \frac{(Sr^{2+}/Ca^{2+})^m_{solid}}{(Sr^{2+}/Ca^{2+})^m_{liquid}} . \tag{VIII.1}$$

The variation of the distribution coefficient with temperature is shown in Fig. VIII.1, which is based on data published by KINSMAN (1969). It is clear that aragonite concentrates Sr^{2+} at temperatures less than about 50°C,

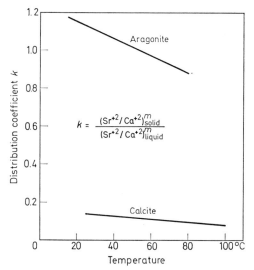

Fig. VIII.1. Temperature dependence of the distribution coefficient for Sr^{2+} in calcite and aragonite precipitating from aqueous solutions. The diagram shows that aragonite concentrates Sr^{2+} relative to Ca^{2+}, when it forms at temperatures of less than about 50° C. Calcite, on the other hand, discriminates against Sr^{2+} at all temperatures within the range of the diagram. Precipitation of calcite from a solution containing Sr^{2+} will therefore increase the Sr/Ca ratio of the aqueous phase, while precipitation of aragonite at temperatures below 50°C will decrease this ratio. (Data taken from KINSMAN, 1969)

while calcite discriminates against it at all temperatures. Consequently, precipitation of calcite will tend to *increase* the Sr/Ca ratio of the oceans, while precipitation of aragonite at low temperatures may *decrease* it. According to data reviewed by KINSMAN (1969), the mole ratio of strontium to calcium in the modern oceans is very nearly constant and has a value of $(0.86 \pm 0.04) \times 10^{-2}$. It is possible that the Sr/Ca ratio of the oceans may have varied in the geologic past depending, at least in part, on the mineralogy of carbonate skeletons and of inorganically-precipitated calcium carbonate. This problem has been considered by several scientists, among them KULP, TUREKIAN, and BOYD (1952), ODUM (1957), LOWENSTAM (1964), and TUREKIAN (1964). So far, it has not been possible to measure the Sr/Ca ratio of the oceans in the geologic past, primarily because carbonate rocks and fossil shells are subject to chemical alteration which changes their Sr/Ca ratios.

The isotopic composition of strontium in ocean water was measured by FAURE, HURLEY, and POWELL (1965), who analyzed surface water from the North Atlantic Ocean and concluded that strontium in this ocean is isotopically homogeneous and has an average $^{87}Sr/^{86}Sr$ ratio of 0.7093 ± 0.0005. More recent analyses of marine strontium from many different locations are

Table VIII.1. Summary of $^{87}Sr/^{86}Sr$ ratios in the oceans

Locality and number of samples	$^{87}Sr/^{86}Sr$	Reference
North Atlantic, Ocean surface water (10)	0.7093 ± 0.0005	FAURE, HURLEY, and POWELL (1965)
Atlantic Ocean, different depths (7)	0.7087 ± 0.0006	HAMILTON (1966)
Hudson Bay, water and shells (8)	0.7093 ± 0.0003	FAURE, CROCKET, and HURLEY (1967)
Ross Sea (2)	0.7094 ± 0.0001	JONES and FAURE (1967)
Seawater, no location given (1)	0.7097 ± 0.0005	BOGARD et al. (1967)
Pacific, Atlantic, and Indian Oceans (23)	0.7094 ± 0.0006	MURTHY and BEISER (1968)
Seawater, no location given (1)	0.7095 ± 0.0004	KAUSHAL and WETHERILL (1969)
Worldwide, mainly shells (9)	0.7090 ± 0.0003	PETERMAN, HEDGE, and TOURTELOT (1970)
Seawater, no location given (1)	0.70910 ± 0.00002	PAPANASTASSIOU, WASSERBURG, and BURNETT (1970)
Red Sea (1)	0.7092 ± 0.0009	IKPEAMA (1971)
Black Sea (1)	0.7093 ± 0.0007	Cox and FAURE (1972)
Composite of modern mollusk shells from different locations (7)	0.70905 ± 0.00006	HILDRETH and HENDERSON (1971)

All errors are one standard deviation. The $^{87}Sr/^{86}Sr$ ratios have been corrected to $^{86}Sr/^{88}Sr = 0.1194$, but interlaboratory discrepancies have not been removed.

compiled in Table VIII.1 after a paper by Cox and FAURE (1972). The results suggest that the isotopic composition of strontium in the oceans is constant. This conclusion is consistent with the long residence time of strontium (1.9×10^7 years) compared to the mixing rate of the oceans (10^3 years). The only exception was reported by FAURE and JONES (1969), who found a $^{87}Sr/^{86}Sr$ ratio of 0.7080 in hot saline brines in the Atlantis II and the Discovery deeps in the median valley of the Red Sea. These brines appear to have been discharged by springs in the bottom of the Atlantis II deep and may contain some strontium of deep-seated origin having a lower $^{87}Sr/^{86}Sr$ ratio than normal marine strontium.

In 1948 the well-known Swedish geochemist WICKMAN published a provocative paper in which he postulated that the isotopic compositions of the elements in seawater may closely approach a crustal average. Specifically he suggested that the decay of ^{87}Rb in the continental crust may have increased the $^{87}Sr/^{86}Sr$ ratio of the oceans at a sufficient rate to make this ratio a useful indicator of time. He pointed out that rubidium is very efficiently excluded from carbonates and sulfates which precipitate in the oceans, making their $^{87}Sr/^{86}Sr$ ratios invariant with respect to time and dependent only on the $^{87}Sr/^{86}Sr$ ratio of seawater at the time of deposition. Using data available at the time, he predicted that the $^{87}Sr/^{86}Sr$ ratios of marine limestones of different geologic age should vary by as much as 6.8 percent per billion years. Such a rate of variation would have made it possible to date marine limestones and sulfate rocks with a high degree of precision. WICKMAN's limestone geochronometer was tested by HERZOG et al. (1953), GERLING and SHUKOLYUKOV (1957), GAST (1955), and HEDGE and WALTHALL (1963). It was found that the variation of $^{87}Sr/^{86}Sr$ ratios of limestones of different ages is significantly less than predicted by WICKMAN. One of several reasons for this is that WICKMAN greatly overestimated the Rb/Sr ratio of the continental crust. In addition, he did not take into consideration the predominant recycling of marine strontium by weathering of marine carbonate rocks and the introduction of primary strontium from the upper mantle as a result of volcanic activity in the ocean basins and along continental margins.

The isotopic composition of strontium in the present-day oceans can be regarded as a mixture of different isotopic varieties of strontium derived from weathering of young volcanic rocks ($^{87}Sr/^{86}Sr \sim 0.704$), marine carbonate rocks ($^{87}Sr/^{86}Sr \sim 0.708$), and old granitic rocks as well as sedimentary rocks derived from them ($^{87}Sr/^{86}Sr \sim 0.715$). It is likely that these sources have contributed varying proportions of the total strontium input into the oceans throughout geologic time. Consequently, it is possible that the $^{87}Sr/^{86}Sr$ ratio in the oceans has varied in the past. PETERMAN, HEDGE, and TOURTELOT (1970) have recently demonstrated that significant changes in the $^{87}Sr/^{86}Sr$ ratio of the oceans did occur in Phanerozoic time. This conclusion is based on analyses of strontium extracted from unreplaced calcium carbonate of fossils. Fig. VIII.2 illustrates their results, which show that the $^{87}Sr/^{86}Sr$ ratio of the oceans during the Mesozoic era was significantly lower than it is now. The lowering of this ratio may have been caused by increased input of strontium derived from volcanic activity which appears to have been unusually widespread during the Mesozoic era. The data of PETERMAN and his colleagues also suggest that remarkable fluctuations of the $^{87}Sr/^{86}Sr$ ratio may have occurred during the Carboniferous and Triassic periods. The significance of these fluctuations has not yet been explained, although ARMSTRONG (1971) suggests that they correlate with the intensity of continental glaciations. The relatively rapid increase of the $^{87}Sr/^{86}Sr$ ratio in the oceans from the Cretaceous period to the Present has been confirmed by DASCH and BISCAYE (1971).

The question regarding the isotopic homogeneity of strontium in the oceans of the geologic past is not yet completely answered. PETERMAN and his colleagues analyzed strontium in fossil shells of Cretaceous age from

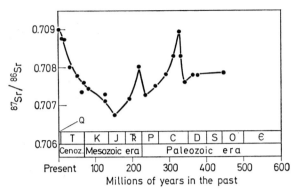

Fig. VIII.2. Variation of the $^{87}Sr/^{86}Sr$ ratio of the oceans during Phanerozoic time. The error bars have been omitted, and a smooth curve has been drawn by us to emphasize the remarkable changes in the $^{87}Sr/^{86}Sr$ ratio of the oceans that are suggested by these data. The most important conclusion is that this ratio was significantly lower during the Mesozoic era than it is now, possibly because of a higher input of strontium from young volcanic rocks at that time. The relatively rapid changes of the $^{87}Sr/^{86}Sr$ ratio during the Triassic and Carboniferous periods have not yet been explained. (Data published by PETERMAN, HEDGE and TOURTELOT, 1970)

many widely-spaced localities in North America and found that the isotopic composition of all samples was the same within experimental error. On the basis of this and similar results for other periods they suggested that the oceans in the past were as well-mixed as they are now. On the other hand, BROOKINS, CHAUDHURI, and DOWLING (1969) analyzed strontium from Permian limestones in eastern Kansas and found variations of the $^{87}Sr/^{86}Sr$ ratio ranging from 0.7070 to 0.7091. They thought it possible that this difference was an indication that local variations could occur in restricted basins of the sea. However, it is also possible that the observed variation is due to removal of radiogenic ^{87}Sr from clay minerals during solution of the limestone samples. FAURE, CROCKET, and HURLEY (1967) reported a normal $^{87}Sr/^{86}Sr$ ratio of 0.7093 ± 0.0003 for water from Hudson Bay, which certainly qualifies as a restricted marine basin. Present indications are, therefore, that strontium in the oceans has been isotopically homogeneous throughout Phanerozoic time, even in shallow basins whose interaction with the open oceans may have been restricted.

While strontium in the oceans appears to be isotopically homogeneous, the strontium of noncarbonate deep-sea sediments has a variable isotopic composition and does not equilibrate with the strontium in the water. DASCH, HILLS, and TUREKIAN (1966), DASCH (1969), and BISCAYE and DASCH (1971) measured $^{87}Sr/^{86}Sr$ ratios of the detrital aluminosilicate fraction of sediment from the Atlantic Ocean. They found important regional

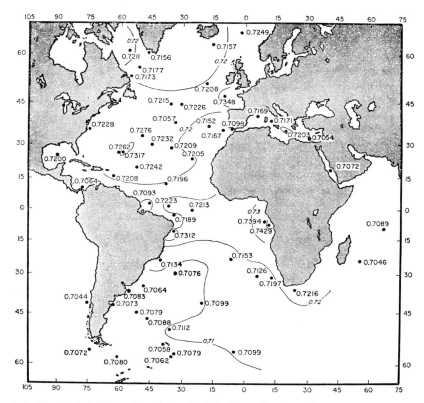

Fig. VIII.3. $^{87}Sr/^{86}Sr$ ratios of the detrital silicate fraction of core-top sediments. The significant regional variation of this ratio in deep-sea sediment indicates the provenance of the detrital silicate minerals derived from adjacent continental areas and emphasizes the fact that strontium does not equilibrate isotopically with sea-water. Reproduced from DASCH (1969) by permission of the executive editor of *Geochimica et Cosmochimica Acta*

variations which reflect the provenance of the sediment. Fig. VIII.3 is a plot of $^{87}Sr/^{86}Sr$ of the silicate fraction of deep-sea sediment in the Atlantic Ocean, with contour lines as drawn by DASCH (1969).

Similar isotope disequilibrium between strontium in sediment and in water was reported by HART and TILTON (1966) for Lake Superior. Their measurements showed that the detrital sediment at the bottom of the lake has an average $^{87}Sr/^{86}Sr$ ratio of about 0.740, while the water has a ratio of only 0.718. Most of the available data suggest that strontium in sediment deposited in the oceans or in lakes does not equilibrate isotopically with strontium in the water. However, an exception to this generalization was reported by JONES and FAURE (1967) for Lake Vanda in Wright Valley of

Southern Victoria Land, Antarctica. Sediment collected from the bottom of this very saline lake was found to have the same $^{87}Sr/^{86}Sr$ ratio as the brine.

The apparent similarity of the isotopic composition of strontium in the brines and sediment of Lake Vanda may not be as exceptional as it at first appears to be. JONES and FAURE (1967) showed that chemical weathering of the rock debris which covers the floor of Wright Valley releases strontium which is isotopically indistinguishable from that remaining in the silicate rock fragments. Since Wright Valley is a closed basin, it is therefore not surprising that the brines of Lake Vanda, the glacial melt-water streams, and the soil in Wright Valley contain strontium of very similar isotopic composition. Isotopic equilibration of strontium between sediment and water is unlikely in an open system such as the Great Lakes of North America or the world ocean.

3. Marine Evaporites

Evaporite rocks occur on many continents in geological systems from the Cambrian to the Quaternary. Most evaporite deposits are of marine origin, although continental evaporites are locally important as sources of lithium, potassium, and boron salts. The literature dealing with evaporite deposits is voluminous and is growing rapidly. Several excellent summaries and texts have been published by BORCHERT and MUIR (1964), BRAITSCH (1962), KRUMMBEIN (1951), and LOTZE (1938). A comprehensive bibliography on evaporite deposits was recently compiled by CRAMER (1969).

Very little is known about the isotopic composition of strontium in evaporite rocks. It is likely that calcium-rich evaporite minerals such as anhydrite or dolomite contain strontium whose isotopic composition is very nearly identical to that of the brine from which they were deposited. Potassium-rich minerals, on the other hand, may contain sufficient rubidium to cause measurable enrichment in radiogenic ^{87}Sr.

Efforts have been made to use potassium-bearing minerals for age determinations by the K-Ar or K-Ca method (GENTNER, PRÄG, and SMITS, 1953; GENTNER, GOEBEL, and PRÄG, 1954; POLEVAYA et al., 1958; PILOT and RÖSLER, 1967; and WARDLAW, 1968). These studies have invariably demonstrated that potassium-bearing evaporite minerals such as sylvite (KCl), carnallite ($KMgCl_3$ 16 H_2O), langbeinite ($K_2Mg_2(SO_4)_3$), and polyhalite ($K_2Ca_2Mg(SO_4)_4 \cdot 2 H_2O$) do not retain the radiogenic ^{40}Ar which is produced by decay of ^{40}K, and consequently they cannot be used to determine the time of deposition. The loss of ^{40}Ar may be continuous or episodic during structural deformation and recrystallization of the minerals.

POLEVAYA et al. (1958) dated sylvite by the K-Ar method as well as by the K-Ca method. They obtained concordant results for a sample of unrecrystallized sylvite, but discordant ages for recrystallized samples. The K-Ca ages were in satisfactory agreement with the known geologic ages of the samples, while the K-Ar ages of recrystallized sylvites were all less than the depositional ages. The authors suggested that radiogenic ^{40}Ca is less mobile than ^{40}Ar and is retained sufficiently well to make sylvite dateable by the K-Ca method. However, no further work has been done to test the feasibility of dating sylvite or potassium-bearing evaporite rocks by the K-Ca method.

The work of POLEVAYA et al. (1958) was concerned only with the applicability of the K-Ar and K-Ca methods of dating. If ^{40}Ca is quantitatively retained in sylvite crystals, it may be worthwhile to investigate the possibility of using the Rb-Sr method of dating of sylvite, because calcium and strontium have similar chemical properties, and ^{87}Sr may therefore be held as firmly as ^{40}Ca appears to be.

It is known that sylvite and carnallite contain appreciable concentrations of rubidium. According to BRAITSCH (1966), sylvites from the evaporite deposits of the upper Rhine Valley have rubidium contents ranging from 57 to 216 ppm. Co-existing carnallites contain in excess of 1000 ppm rubidium. WARDLAW (1968) reported that red-stained sylvite of the Prairie Evaporite Formation of Saskatchewan, Canada, contains from 10 to 100 ppm rubidium, while carnallite contains from 50 to 350 ppm. Little is known about the strontium content of sylvite or carnallite, but it is likely to be less than 10 ppm, except in samples containing accessory anhydrite or calcite. This brief sampling of the geochemical literature suggests that potassium-bearing evaporite rocks of Phanerozoic age may have sufficiently large Rb-Sr ratios to cause a measurable increase of their $^{87}Sr/^{86}Sr$ ratios. If the radiogenic ^{87}Sr is quantitatively retained in whole-rock specimens of evaporite rock containing sylvite or carnallite, the whole-rock Rb-Sr isochron method may be useful for dating of marine evaporite deposits. This method of dating evaporites has not yet been tested.

4. Dating of Detrital Sedimentary Rocks

It is well-established that suites of comagmatic igneous rocks having simple geologic histories satisfy the assumptions of the Rb-Sr method of dating. Specifically, such rocks initially all have the same $^{87}Sr/^{86}Sr$ ratio, become closed systems at very nearly the same time, and do not lose or gain rubidium or strontium. We showed in Chapter II how the common age of such a suite of rocks can be determined from the slope of their isochron.

While it is plausible that rocks crystallizing from a magma should satisfy the conditions for dating listed in the preceding paragraph, it seems unlikely that detrital sedimentary rocks would also satisfy all of these conditions. Nevertheless, COMPSTON and PIDGEON (1962) and WHITNEY and HURLEY (1964) independently discovered that samples of Paleozoic shale generated remarkably good straight lines on isochron diagrams. The dates calculated from these isochrons were either in agreement with the known ages of the rocks, or only slightly greater. Apparently, the strontium of fine-grained detrital sedimentary rocks can approach isotopic homogeneity at some time in the history of the rocks, which then become closed to rubidium and strontium, satisfying the conditions for dating by the Rb-Sr method. DASCH (1969) has discussed the mechanisms leading to isotopic homogenization of strontium in sedimentary rocks deposited in different geological environments. Many isochrons for sedimentary rocks have been reported following the original discovery, and new age determinations of sedimentary rocks are being published each year. A sampling of the literature includes papers by ALLSOPP and KOLBE (1965), PETERMAN (1966), CHAUDHURI and FAURE (1967), OBRADOVICH and PETERMAN (1968), FAURE, MURTAUGH, and MONTIGNY (1968), ALLSOPP, ULRYCH, and NICOLAYSEN (1968), FAIRBAIRN et al. (1969), MOORBATH (1969), POWELL, SKINNER, and WALKER (1969), FAURE and KOVACH (1969), CHAUDHURI and BROOKINS (1969), and BOFINGER, COMPSTON, and GULSON (1970).

Fig. VIII.4 shows an isochron for shale samples of the Fig Tree Series of South Africa reported by ALLSOPP, ULRYCH, and NICOLAYSEN (1968). The slope of this isochron is consistent with a date of 2.98 $\pm$ 0.02 billion years. These rocks are among the oldest fossiliferous sedimentary rocks known. The goodness of fit of the shale samples on the isochron indicates a high degree of isotopic homogenization, which may have resulted from either diagenesis or subsequent metamorphism. Such uncertainty regarding the cause of isotopic homogenization of strontium in sedimentary rocks makes

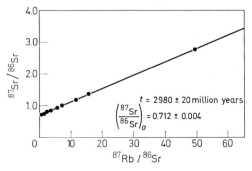

Fig. VIII.4. Rubidium-strontium isochron for samples of shale from the Fig Tree Series of South Africa. The Fig Tree Shale is one of the oldest fossiliferous sedimentary rocks known. The "goodness of fit" of the shale samples indicates that these rocks achieved a high degree of isotopic homogeneity at some time in their history. This time may have been the deposition and diagenesis of the Fig Tree Shale or its subsequent metamorphism. (Data taken from ALLSOPP, ULRYCH, and NICOLAYSEN, 1968)

interpretation of their apparent isochron ages difficult. The antiquity of these sedimentary rocks is of great scientific interest because their very existence indicates that water was present on the surface of the Earth perhaps four billion years ago, as suggested by DONN, DONN, and VALENTINE (1965). Moreover, chert from the Fig Tree Series and the underlying Onverwacht Series contain alga-like fossils and well-preserved bacteria testifying to the fact that primitive forms of life existed on the Earth about 3 billion years ago (BARGHOORN and SCHOPF, 1966; and ENGEL et al., 1968).

5. Glauconite

A discussion of the isotopic composition of strontium in sedimentary rocks is not complete without consideration of the mineral glauconite. Glauconite is a dioctahedral mica containing potassium, iron, aluminum, and magnesium as essential components, with excess water. It occurs almost exclusively in marine sedimentary rocks ranging in age from Precambrian to Recent, and is believed to be authigenic. The occurrence and mineralogy of glauconite have been dealt with by CLOUD (1955) and BURST (1958), among others. It commonly occurs as small green pellets consisting of aggregates of extremely small crystals which are about 1 micron in size, according to GRIM (1952). Because of its high potassium content and authigenic origin, glauconite has been thoroughly investigated to determine its reliability for dating by the K-Ar method. The history and conclusions of these efforts have been reviewed by POLEVAYA, MURINA, and KAZAKOV (1961), KAZAKOV (1964), and HURLEY (1966) who in addition compiled an extensive list of references to the literature.

The problem with glauconite is that it does not satisfy the "closed system" assumption required for dating. K-Ar dates of glauconites therefore are commonly too low by about 10 to 20 percent. Laboratory experiments by EVERNDEN et al. (1960) and POLEVAYA, MURINA, and KAZAKOV (1961) show that radiogenic ^{40}Ar is lost at 200° C and that appreciable losses may occur at temperatures as low as 100° C, if this temperature is maintained for periods of a few million years. One reason for the apparent loss of radiogenic ^{40}Ar appears to be the progressive recrystallization of glauconite, accompanied by a decrease in the percentage of expandable layers, in response to increasing pressure and temperature during burial (HURLEY, 1966). During this process, radiogenic ^{40}Ar may be lost or potassium may be gained, or both, causing K-Ar ages of glauconite to be somewhat lower than mica ages in related igneous rocks. The apparent deficiency of ^{40}Ar increases with depth of burial, degree of metamorphism, and tectonic deformation. Nevertheless, K-Ar dates of glauconite give useful minimum estimates of the time of deposition of sedimentary rocks.

Glauconite may contain several percent potassium, but only a fraction of a percent of calcium. Consequently it is to be expected that glauconite also concentrates rubidium, but excludes strontium, making it potentially suitable for dating by the Rb-Sr method. The first attempt to date glauconite by this method was made by CORMIER (1957) and HERZOG, PINSON, and CORMIER (1958). They found that the glauconite samples which they analyzed contained from 185 to 310 ppm rubidium, but generally less than 20 ppm strontium. The dates calculated for these glauconites were found to be in reasonable agreement with K-Ar dates of other glauconites published previously by other investigators. The general conclusion was that dating of glauconite by the Rb-Sr method was technically feasible and that the dates were consistent with the geologic time scale published by HOLMES in 1947.

Since then, glauconites have been dated by other investigators using the Rb-Sr method with results which have inspired cautious optimism (HURLEY et al., 1960; GULBRANDSEN, GOLDICH, and THOMAS, 1963; McDOUGALL et al., 1965; OBRADOVICH and PETERMAN, 1968). It is difficult to be certain about the retention of radiogenic ^{87}Sr in glauconite because the time of deposition of the sedimentary rocks in which the mineral occurs is commonly not sufficiently well-defined by other dating methods. In general, Rb-Sr as well as K-Ar dates of glauconite tend to underestimate the depositional age of sedimentary rocks, but approach it in cases where the glauconite has not been extensively recrystallized as a result of deep burial, tectonic deformation, or metamorphism. No Rb-Sr age determinations of glauconite have been reported which clearly exceed the depositional age of the sediment. This suggests that inherited excess radiogenic ^{87}Sr is not a problem, although such cases could conceivably be due also to redeposition of glauconite derived from older rocks.

OBRADOVICH and PETERMAN (1968) have made a very interesting study of the age of the Precambrian Belt Series of Montana by dating glauconite as well as argillaceous sedimentary rocks by the Rb-Sr isochron method. Their data for the Sun River sequence of the Belt Series are shown in Fig. VIII.5. It is interesting to note that in this case the glauconites give essentially the same date as the argillaceous sediment samples, although the latter had a significantly higher initial ratio. The pooled isochrons result in a best estimate of 1097 $\pm$ 20 million years for the age of these rocks. The initial ratio of the sediment isochron is 0.7325, and results from the presence of inherited radiogenic ^{87}Sr which, however, does not affect the apparent age derived from the isochron. K-Ar dates of the glauconites are in satisfactory agreement with the Rb-Sr isochron, and both agree with the date indicated by the sediment isochron. According to OBRADOVICH and PETERMAN (1968), the concordance of the results indicates that the event being dated is either (1) deposition or diagenesis, or (2) the subsequent metamorphism and isotopic homogenization of both the rock and mineral systems. Because of the presence of the poorly-ordered 1 M_d mica polymorph (MAXWELL and

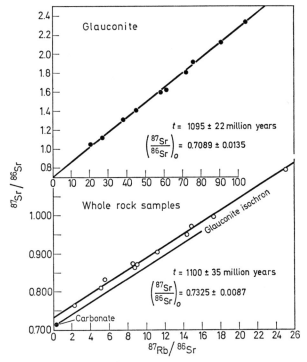

Fig. VIII.5. Rb-Sr isochron of glauconite and argillite. Glauconites and whole-rock samples of argillaceous sedimentary rocks from the Sun River Sequence of the Belt Series in Montana have concordant Rb-Sr isochron ages, although their initial $^{87}Sr/^{86}Sr$ ratios differ significantly. OBRADOVICH and PETERMAN, who analyzed these rocks and minerals, concluded from this and other evidence that these isochrons date the time of deposition of this part of the Belt Series at 1097 ± 20 million years ago. A dolomitic carbonate from the Helena Formation fits the glauconite isochron, consistent with the assumed authigenic origin of this mineral. (Data taken from OBRADOVICH and PETERMAN, 1968)

HOWER, 1967) and on the basis of other age determinations on the sills intruding the Belt and Purcell rocks in Alberta and Montana and on the dates for hornfels produced by the Purcell lavas, the authors concluded that the date indicated by the pooled isochrons is the time of deposition of the Sun River sequence of the Belt Series. This conclusion implies that, at least in this instance, the glauconite has not lost radiogenic ^{40}Ar or ^{87}Sr and that the Rb-Sr isochron for the argillaceous rocks indicates the time of deposition rather than subsequent metamorphism.

Another important result of the above study is that a sample of lime-stone from the Helena Dolomite of the Belt Series plots on the glauconite isochron, as shown in Fig. VIII.5. Apparently the glauconites incorporated strontium from the water in which they were formed, consistent with the assumption that they are authigenic. It also indicates that the initial $^{87}Sr/^{86}Sr$ ratio of the glauconite isochron is indicative of the isotopic composition of seawater at the time of Belt sedimentation. The glauconite isochron therefore provides a measurement of the isotopic composition of strontium in the oceans during Late Precambrian time, assuming that the environment of deposition was in fact marine and provided that the Precambrian oceans were as homogenous isotopically as the modern oceans appear to be.

6. Nonmarine Carbonate Rocks

The isotopic composition of nonmarine carbonate rocks is likely to be variable, depending on the ages and Rb/Sr ratios of the rocks in the watershed which contributed dissolved strontium to the lake in which deposition occurred. FAURE, HURLEY, and FAIRBAIRN (1963) found that lakes and rivers on the Canadian Precambrian Shield have $^{87}Sr/^{86}Sr$ ratios ranging from 0.712 to 0.726 in accord with the antiquity of the bedrock in that area. Most of the measurements were made by analyzing strontium in the shells of mollusks which inhabit the lakes and rivers and which concentrate strontium in their aragonitic shells. If carbonate rocks are formed in lakes such as those on the Canadian Precambrian Shield, their strontium should be variably enriched in radiogenic ^{87}Sr compared to contemporary seawater. Similarly, if the drainage basin of a lake is underlain predominantly by young volcanic rocks, calcium carbonate forming in it should have a low $^{87}Sr/^{86}Sr$ ratio, similar to that of the volcanic rocks. We suggest, therefore, that the $^{87}Sr/^{86}Sr$ ratio of the carbonate minerals in nonmarine carbonate rocks may be a useful parameter for the study of such rocks.

This possibility has been investigated by BARRETT and FAURE (1972), who measured $^{87}Sr/^{86}Sr$ ratios of calcareous rocks of Devonian and Permian ages from the nonmarine Beacon Supergroup of the Transantarctic Mountains. They found that the carbonate phases in these rocks have $^{87}Sr/^{86}Sr$ ratios ranging from 0.715 to 0.729 and are therefore strongly enriched in radiogenic ^{87}Sr, compared to the oceans in those periods of geologic time. BARRETT and FAURE suggested that the apparent enrichment in radiogenic ^{87}Sr of these fresh-water carbonates could be used to obtain information about changes in the provenance of the strontium and to set limits on the ages of the bedrock in the drainage basins from which the strontium was derived.

The isotopic composition of strontium in nonmarine carbonate rocks deserves more attention than it has received in the past.

7. Summary

The isotopic composition of strontium in the oceans is apparently uniform, but it changed systematically during Phanerozoic time. The best estimate of the $^{87}Sr/^{86}Sr$ ratio of the oceans today is 0.7091. The lowest values occurred during the Jurassic period, when the $^{87}Sr/^{86}Sr$ ratio was 0.7067. The variation may be due to increased input of strontium derived by weathering of young volcanic rocks during the Mesozoic era.

Potassium-bearing marine evaporite rocks probably have sufficiently high Rb/Sr ratios to cause measurable enrichment of such rocks in radiogenic ^{87}Sr. However, the retentivity of such rocks for strontium and rubidium has not yet been investigated.

Unmetamorphosed shales and siltstones commonly form isochrons which may indicate the amount of time elapsed since deposition and diagenesis. However, even low-grade regional metamorphism may cause isotopic re-equilibration or loss of ^{87}Sr, leading to an underestimation of the age. This method is most useful for dating unmetamorphosed sedimentary rocks of Precambrian age which usually cannot be dated by any other known methods.

Glauconite is useful for dating, especially by the Rb-Sr isochron method, provided the rock in which it occurs has not been deeply buried, structurally deformed, or metamorphosed.

Nonmarine carbonate rocks may contain strontium whose isotopic composition reflects the age and Rb/Sr ratio of bedrock in the watershed which provided dissolved strontium to the depositional basin.

IX. Isotopic Homogenization of Strontium in Open Systems

1. Introduction

After the presentation and discussion of the isotopic composition of strontium in different igneous and sedimentary rocks in the preceding chapters, we now return to the problem of dating by the Rb-Sr method. We stated in Chapter II that one of the prerequisites for dating rocks or minerals is that their concentrations of rubidium and strontium and their $^{87}Sr/^{86}Sr$ ratios must have changed only as a result of decay of ^{87}Rb to ^{87}Sr. When this condition is not satisfied, the dates calculated from the equations derived in Chapter II do not indicate the age of a rock, but may instead date the end of the alteration process that disturbed its ^{87}Rb-^{87}Sr decay scheme.

The interpretation of dates derived from rocks and minerals which have been altered is complicated for several reasons. First of all, there may be no mineralogical or textural evidence to warn the geochronologist that an igneous rock or any of its minerals he is analyzing for an age determination have been altered. Secondly, the disturbance of the ^{87}Rb-^{87}Sr decay schemes may have affected only the minerals, while the total rock remained a closed system. The alteration of the decay schemes in the minerals of a rock generally results in discordant dates, none of which may have geological meaning. However, the discordance of mineral dates is a criterion that identifies rocks that have been altered.

It is usually not possible to identify the specific mechanism which is responsible for the failure of a mineral or rock to satisfy the closed-system requirement. It may involve no more than the exchange of radiogenic ^{87}Sr with other strontium isotopes in the pore fluid or the loss of the radiogenic isotope by diffusion out of the crystal lattice in which it formed. On the other hand, it may also be due to metasomatism resulting in gain or loss of rubidium and/or strontium by minerals, or of the whole rock. Fortunately, it is possible in many cases to interpret the observed ^{87}Rb-^{87}Sr decay schemes in altered rocks and minerals without knowing the exact cause of the alteration.

The geological processes which may affect the ^{87}Rb-^{87}Sr decay schemes of rocks and minerals primarily include metamorphism (both contact and regional) and chemical weathering. The actions of hydrothermal fluids or hot intrastratal brines on the isotopic composition and concentrations of strontium and rubidium are undoubtedly equally profound. However, in the presentation which follows we shall discuss only the effects of contact and regional metamorphism and of chemical weathering, because good examples of these have been described in the literature and because they will permit us to explain the techniques that have been developed for the interpretation of dates derived from altered rocks and minerals.

The systematics of isotopic homogenization of strontium among minerals and possible concurrent migration of rubidium and strontium within crystalline rocks have been discussed by COMPSTON, JEFFERY, and RILEY (1960), COMPSTON and JEFFERY (1961), NICOLAYSEN (1961), FAIRBAIRN, HURLEY, and PINSON (1961), RILEY and COMPSTON (1962), LANPHERE et al. (1964), LONG (1964) ALLEGRE and DARS (1965), and ARRIENS et al. (1966). Many other geochronologists, too numerous to mention here, have contributed observations and interpretations of isotopic systems disturbed by metamorphism.

2. Contact Metamorphism

The dominant parameter which determines to what extent the ^{87}Rb-^{87}Sr decay scheme in minerals is changed is the temperature. The effect of elevated temperature on the retentivities of the minerals that are commonly dated by the Rb-Sr method can be studied under laboratory conditions. However, efforts to study the migration of isotopes of rubidium and strontium between minerals under laboratory conditions have been much less successful than similar studies regarding the diffusion of radiogenic ^{40}Ar (AMIRKHANOV, BRANDT, and BARTNITSKII, 1959, 1961; EVERNDEN et al., 1960; and FECHTIG and KALBITZER, 1966). Fortunately there is another way to study this migration. Rocks from contact metamorphic aureoles provide an opportunity for investigating the loss of radiogenic daughters from a variety of minerals whose thermal histories can be approximated by suitable heat-flow models. Such studies have an additional advantage over laboratory studies because enough time is available in nature for sluggish reactions to reach equilibrium. The effect of contact metamorphism on mineral dates has been investigated by HART (1964), HANSON and GAST (1967), and HART et al. (1968). The presentation that follows is based primarily on the work of Hart and his colleagues, who made a very complete and thorough study of the effects of contact metamorphism on the Eldora Stock on the Idaho Springs Formation in the Front Ranges of Colorado.

The Idaho Springs Formation consists of highly-metamorphosed sedimentary rocks, with minor volcanic rocks of Precambrian age. The rocks are quartz-feldspar-biotite gneisses with some amphibolite and abundant, but small, bodies of pegmatite. The metamorphic grade is sillimanite-almandine subfacies of the almandine amphibolite facies. The Idaho Springs

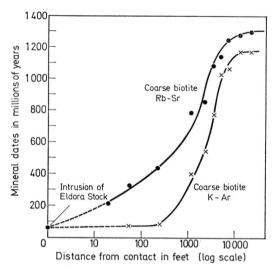

Fig. IX.1. Variation of Rb-Sr and K-Ar dates of coarse-grained biotite in the contact metamorphic aureole of the Eldora Stock in the Idaho Springs Formation, Front Range, Colorado. The dates reflect varying losses of radiogenic ^{87}Sr and ^{40}Ar, which are related to the maximum temperature and duration of heating of the country rock. (Data from HART, 1964)

Formation is cut by large granitic intrusives and pegmatites of Precambrian age and by a series of granitic stocks of Tertiary age. Contact metamorphism associated with the intrusion of these stocks caused subtle mineralogical changes in the rocks of the Idaho Springs Formation but very profound changes in their isotopic systems. The mineralogical changes were described by HART (1964) and STEIGER and HART (1967). The effects on lead isotopes in potassium feldspar were studied by DOE and HART (1963), while TILTON et al. (1964) and HART et al. (1968) reported changes in the U-Pb and Th-Pb decay schemes in zircon from the same rocks.

Fig. IX.1 shows the systematic variation of Rb-Sr and K-Ar dates of coarse biotite of the Idaho Springs Formation as a function of distance from the contact with the Eldora Stock, one of the Tertiary intrusives mentioned above. The diagram shows that the Rb-Sr dates of coarse biotites increase smoothly away from the contact to a distance of 14100 feet, where they reach a value of 1275 million years. This date probably refers to an earlier Precambrian metamorphism of the Idaho Springs Formation, which may have been deposited prior to 1600 million years ago, according to a review of pertinent age determinations by HART (1964). The age of the Eldora Stock and therefore of the most recent contact metamorphism is 54 million years.

The pattern of Rb-Sr and K-Ar dates of biotites from the contact aureole of the Eldora Stock can be explained in terms of variable loss of radiogenic ^{87}Sr and ^{40}Ar in response to an increase in temperature associated with the contact metamorphism. HART (1964) calculated that biotite 20 feet from the contact had lost 88 percent of radiogenic ^{87}Sr, while the loss at 14100 feet was essentially zero. The pattern of radiogenic ^{40}Ar loss from the biotites is similar, except that the fraction lost is somewhat higher.

The loss of radiogenic ^{87}Sr from a mineral may be due to diffusion alone, or it may be controlled by the kinetics of a chemical reaction occurring in the rock. In either case, the fraction of daughter lost is related to the maximum temperature to which the mineral was subjected and to the period of time during which the process was effective. The time-dependent variation of the temperature in the country rock can be approximated from heat-flow theory (JAEGER, 1957, 1959; LOVERING, 1935; CARSLAW and JAEGER, 1959). Assuming spherical diffusion, the fraction of daughter lost (F) can then be used to calculate appropriate values of D/a^2, where D is the diffusion coefficient and a is the radius of a hypothetical sphere out of which diffusion has occurred. The diffusion coefficient is temperature-dependent in the form:

$$\frac{D}{a^2} = \frac{D_0}{a^2} e^{-Q/RT}, \tag{IX.1}$$

where Q is the "activation energy" required to dislodge and move a ^{87}Sr atom from one stable lattice site to another, R is the universal gas constant, and T is the absolute temperature in degrees Kelvin. The value of the activation energy associated with the diffusion of radiogenic ^{87}Sr can be determined graphically. Taking logarithms to the base 10 of Eq. (IX.1) we obtain:

$$\log\left(\frac{D}{a^2}\right) = \log\left(\frac{D_0}{a^2}\right) - \frac{Q}{2.303\,RT}, \tag{IX.2}$$

which is the equation of a family of straight lines in coordinates of $\log(D/a^2)$ and $(1/T)$. By plotting the logarithm of D/a^2 calculated for each biotite versus the reciprocal of the estimated temperature, the value of the activation energy can be derived from the slope of the resulting straight line. HART (1964) obtained values ranging from 16 to 27 kcal/mole, depending on the assumed geometry of the heat-flow models.

Several attempts have been made to study thermally-induced migration of radiogenic ^{87}Sr, rubidium, and strontium in minerals under laboratory conditions. The most successful of these was reported by BAADSGAARD and VAN BREEMEN (1970). These authors heated specimens of the Prosperous Lake Granite of the Northwest Territories of Canada (actually the rock is a quartz monzonite) in air for 100 hours at temperatures up to 1025°C. Subsequent analyses of separated minerals from the heated specimens showed that significant changes had occurred in the $^{87}Sr/^{86}Sr$ ratios and the concentrations of rubidium and strontium of the minerals, while the whole rock had remained closed. The $^{87}Sr/^{86}Sr$ ratios of all of the minerals changed in the direction of the $^{87}Sr/^{86}Sr$ ratio of the rock, as shown in Fig. IX.2. The results of this important experiment suggest that heating may result in isotopic homogenization of strontium and cause changes in the concentrations of rubidium and strontium of the minerals.

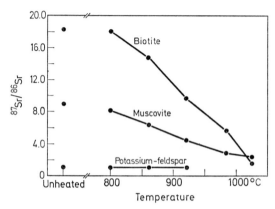

Fig. IX.2. Change in the $^{87}Sr/^{86}Sr$ ratios of biotite, muscovite, and potassium -feldspar in a specimen of quartz monzonite, after heating for 100 hours at each of the temperatures indicated on the diagram. The graphs emphasize the fact that the $^{87}Sr/^{86}Sr$ ratios of these minerals were lowered as the result of heating and approach the ratio of the total-rock ($^{87}Sr/^{86}Sr = 1.079$), which remained unaffected even after heating at 1025°C. (Data from BAADSGAARD and VAN BREEMEN, 1970)

The preceding example of the effects of heating on the ^{87}Rb-^{87}Sr decay schemes of minerals indicates that the $^{87}Sr/^{86}Sr$ ratios of rubidium-rich minerals are lowered, while those of strontium-rich minerals are increased. If this process goes to completion, the $^{87}Sr/^{86}Sr$ ratios of all the minerals may be equalized to the ratio of the total rock, which remains a closed system. The effects of isotopic homogenization can be illustrated on a strontium development diagram, which we introduced in Chapter II.

Let us consider a hand specimen of a granitic rock containing biotite, potassium-feldspar, and a strontium-rich phase such as apatite. Fig. IX.3 illustrates the changes in the $^{87}Sr/^{86}Sr$ ratios that occur in the minerals during an episode of metamorphism. We assume that after initial crystallization and cooling at t_1, this rock was reheated for a period of time $\Delta t = t_4 - t_5$, followed by cooling to a low ambient temperature at t_5. During this episode the $^{87}Sr/^{86}Sr$ ratios of all of the minerals were homogenized, but the whole-rock remained a closed system.

In this hypothetical example the $^{87}Sr/^{86}Sr$ ratios of biotite and potassium-feldspar are both lowered while that of apatite is increased until all the minerals have the same $^{87}Sr/^{86}Sr$ ratio as the whole rock. As a result, the strontium development lines of the minerals converge to a point at t_5, which is the time elapsed since the minerals became closed systems following iso-topic re-equilibration. If dates were calculated for each mineral by assuming an initial $^{87}Sr/^{86}Sr$ ratio, the dates would be *discordant*. Fig. IX.3 shows the dates that would result for a particular assumed value of the initial ratio, here taken to be equal to that of the whole rock at the time of crystallization. The date calculated for the potassium-feldspar would be t_2, while that for the biotite would be t_3. Both dates are *fictitious* and have no geologic significance. The only meaningful date that can be derived from the minerals in Fig. IX.3 is t_5, which dates the end of the metamorphic episode and the renewed accumulation of radiogenic ^{87}Sr in the minerals. In conclusion we emphasize that the isotopic composition of strontium in the whole rock is unaffected by metamorphism in the hypothetical case shown in Fig. IX.3. COMPSTON and JEFFERY (1959), SCHREINER (1958), and NICOLAYSEN (1961) first reported actual examples of this important fact.

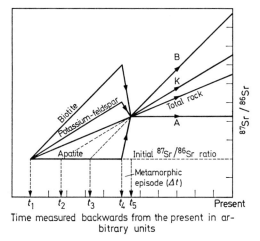

Fig. IX.3. Strontium development diagram showing isotopic homogenization of strontium in the minerals of a granitic rock during metamorphism. In this hypo-thetical case, the time interval during which homogenization occurred has been greatly exaggerated to emphasize that this is not an instantaneous process even on a geologic time scale. The diagram assumes that the Rb/Sr ratios of the minerals remained unchanged and that the total-rock remained closed during the meta-morphic episode. Dates calculated from the minerals using an assumed initial $^{87}Sr/^{86}Sr$ ratio are discordant, as suggested by the dashed lines. The only geological-ly significant date derivable from the data in the diagram is t_5, the time elapsed since the end of metamorphism. (After FAIRBAIRN, HURLEY, and PINSON, 1961)

Discordance of mineral dates of granitic rocks with metamorphic histories are the rule rather than the exception. Examples of this phenomenon have been reported and discussed by many geochronologists, including WETHERILL, DAVIS, and TILTON (1960), ALLSOPP (1961), GILETTI, MOORBATH, and LAMBERT (1961), ALLSOPP et al. (1962), DEUSER and HERZOG (1962), JÄGER (1962), LONG and LAMBERT (1963), FAURE et al. (1964), GRANT (1964), ALDRICH, DAVIS, and JAMES (1965), WASSERBURG and LANPHERE (1965), WETHERILL and BICKFORD (1965), BROOKS (1966), WHITE, COMPSTON, and KLEEMAN (1967), CAHEN, DELHAL, and DEUTSCH (1967), PETERMAN, HEDGE, and BRADDOCK (1968), BURCHART (1968), WANLESS, LOVERIDGE, and MURSKY (1968), JÄGER and ZWART (1968), HANSON, GRÜNENFELDER, and SOPTRAYANOVA (1969), BICKFORD et al. (1969), KROGH and DAVIS (1969), HEIER and COMPSTON (1969), and CAHEN et al. (1970).

3. Regional Metamorphism

We shall now present an example of the effects of regional metamorphism on a suite of granitic rocks from the Baltimore Gneiss that was studied by WETHERILL et al. (1966) and WETHERILL, DAVIS, and LEE-HU (1968). Actually, these effects are similar in principle to those just discussed. The main difference is that regional metamorphism occurs on a larger scale, may act over a longer interval of time, and consequently more often leads to complete isotopic re-equilibration of minerals.

The Baltimore Gneiss occurs in the form of domes which are mantled by Late Precambrian metasedimentary rocks of the Glenarm Series. Both units were intruded by younger magmas, and deformed and metamorphosed during part of the Paleozoic Appalachian orogeny between 600 and 400 million years ago. WETHERILL, DAVIS, and LEE-HU (1968) analyzed total rocks and separated mineral fractions from the Baltimore Gneiss. Their results are shown as an isochron diagram in Fig. IX.4.

Four of the five total rock samples define an isochron whose slope indicates a date of 1050 ± 100 million years and an initial $^{87}Sr/^{86}Sr$ ratio of 0.705 ± 0.002. This date is identical to nearly concordant U-Pb dates on zircons from this rock reported previously by TILTON et al. (1958), and very probably represents the geologic age of these rocks. Biotite, potassium-feldspar, and plagioclase of two of the five total-rock samples form distinctly different isochrons. Their slopes are virtually identical and much lower than the slope of the total-rock isochron. They indicate that these minerals became closed systems only about 290 million years ago. This date may reflect the slow cooling of these rocks, following the magmatic activity and regional metamorphism associated with the Appalachian orogeny. The initial $^{87}Sr/^{86}Sr$ ratios of the two mineral isochrons are 0.745 and 0.776,

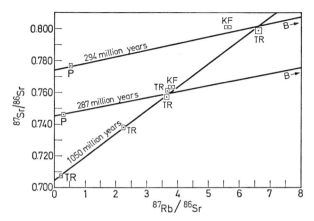

Fig. IX.4. Whole-rock and mineral isochrons for the Baltimore Gneiss. The whole-rock samples define an isochron whose slope is clearly greater than that of two mineral isochrons. The interpretation of these data is that the Baltimore Gneiss originally crystallized 1050 million years ago and was subsequently metamorphosed during the Appalachian orogeny. As a result, the strontium in the minerals was isotopically homogenized by exchange of ^{87}Sr and by possible migration of rubidium and strontium. The minerals became closed systems about 290 million years ago and have remained closed since that date. (Data from WETHERILL, DAVIS, and LEE-HU, 1968)

respectively. Both are significantly greater than the initial ^{87}Sr/^{86}Sr ratio of the total rocks.

The Baltimore Gneiss apparently experienced internal re-equilibration of strontium in its minerals during regional metamorphism, while total-rock samples remained closed systems. We shall next present a hypothetical example of this phenomenon and trace the movement of total-rock samples and their constituent minerals on an isochron diagram.

Fig. IX.5 is an isochron diagram showing the effects of internal isotopic re-equilibration on a suite of three comagmatic rocks (TR1, 2, and 3) and three minerals of TR2: biotite (B2), potassium-feldspar (KF2), and apatite (A2). Initially, the rock specimens and their minerals had the same ^{87}Sr/^{86}Sr ratio, forming an isochron with a slope of zero. Subsequently the points representing each total rock and each mineral moved along trajectories with a slope equaling -1 during the time interval τ_1. After that, the ^{87}Sr/^{86}Sr ratios of the minerals were changed during a short episode of metamorphism. The arrows in Fig. IX.5 show the ^{87}Sr/^{86}Sr ratios of B2 and KF2 were lowered while that of A2 was raised until all the minerals had the same ^{87}Sr/^{86}Sr ratio as TR2. The ^{87}Rb/^{86}Sr ratios are assumed to have remained constant. At the end of the episode of metamorphism the minerals formed a new isochron (indicated by the dashed horizontal line) having a slope of zero. The minerals of the other rocks were similarly re-equilibrated, but they

7*

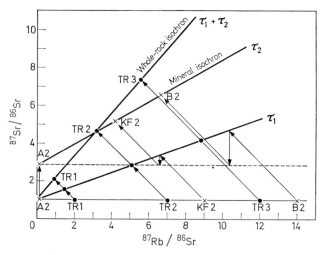

Fig. IX.5. This diagram traces the paths taken by three comagmatic igneous rocks (TR 1, 2, and 3) and the minerals of TR 2 which were isotopically homogenized during an episode of metamorphism τ_1 years after initial crystallization. The movement of the rocks and minerals on this isochron diagram is described in the text. This hypothetical case is analogous to the history of the Baltimore Gneiss shown in Fig. IX.4

are not shown in order to preserve the clarity of the diagram. The total-rock samples were unaffected by the isotopic homogenization of strontium in the minerals and therefore continued on their original trajectories. At the end of the short metamorphic episode all minerals were aligned on their own isochron, whose slope increased by an appropriate amount in the time interval τ_2, which brings us to the Present. The slope of the mineral isochron formed by B2, KF2, TR2, and A2 will yield τ_2, the time elapsed since they became closed systems following the last episode of metamorphism. The minerals of the other rocks would form parallel isochrons which would yield the same or similar values for τ_2, but would have different initial $^{87}Sr/^{86}Sr$ ratios. The mineral isochrons therefore date the end of the metamorphic episode, when the temperature dropped sufficiently to permit radiogenic ^{87}Sr to accumulate in the minerals. The whole rocks, on the other hand, form a separate isochron whose slope measures the total amount of time $(\tau_1 + \tau_2)$ which has elapsed since initial crystallization of the suite of rocks. The whole-rock isochron also indicates the value of the initial $^{87}Sr/^{86}Sr$ ratio of these rocks.

4. Open-System Behavior of Total-Rock Samples

So far we have considered only cases in which minerals were isotopically re-equilibrated or chemically altered while total-rock samples remained unaffected. Large-scale re-equilibration of large bodies of igneous rocks appears to be a rare occurrence, but may be common in metamorphosed sedimentary rocks.

WASSERBURG, ALBEE, and LANPHERE (1964) have reported a spectacular example of apparent enrichment of total rock samples in radiogenic ^{87}Sr during metamorphism. They studied hornblende diorite dikes from the Panamint Mountains of California. These dikes intrude Precambrian augen gneiss and cross-cutting granites of the World Beater Complex, as well as late Precambrian metasediments of the Pahrump Series which rest unconformably on the older granites and gneisses. The age of this "Pahrump diabase" is believed to be between 600 and 1350 million years. Earlier work by LANPHERE et al. (1964) showed that minerals of the augen gneiss and the granite of the World Beater Complex form isochrons corresponding to dates of 100 to 150 million years. However, the total-rock samples scattered rather widely and did not fit an isochron. The geologic ages of these rocks were determined by the U-Pb method using zircon and were found to be 1800 million years for the augen gneiss and 1350 million years for the granite. These rocks therefore appear to have become open systems as a result of metamorphism during Mesozoic time.

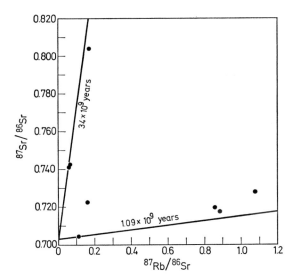

Fig. IX.6. Whole-rock isochron diagram for Pharump diabase from the Panamint Mountains. The points representing whole-rock samples scatter widely, which indicates that they have not remained closed systems. In this case the scatter of the points can be attributed to the addition of varying amounts of radiogenic ^{87}Sr derived from the Precambrian gneiss and granite of the World Beater Complex during Mesozoic metamorphism. (Data from WASSERBURG, ALBEE, and LANPHERE, 1964)

The analytical results for total-rock samples of the "Pahrump diabase" reported by WASSERBURG, ALBEE, and LANPHERE are shown in Fig. IX.6. It is readily apparent that these rocks scatter widely on the isochron diagram. Two reference isochrons have been drawn to show that dates ranging from 1.09 to 34 billion years could be calculated for individual specimens relative to an initial ratio of 0.703. Dates in excess of the age of the earth (4.6×10^9 years) are obviously not acceptable. A possible explanation for the scatter of points on the isochron diagram is that these rocks may have been variously enriched in radiogenic ^{87}Sr which might have been derived from the adjacent granite and gneiss during Mesozoic metamorphism. These results indicate that even total-rock systems may be open during metamorphism and may have their isotopic systems changed, making it impossible to determine their geologic age.

In summary, meaningful dates can be derived from altered rocks under the following conditions: (1) if isotopic homogenization has occurred among the minerals of a rock, the mineral isochron indicates the time elapsed since re-equilibration; (2) if total-rock samples remained closed systems during the re-equilibration of the minerals, the total-rock isochron gives the time elapsed since crystallization and thus the "age" of the rocks; (3) if the total rocks were open to rubidium and strontium, but the minerals were isotopically homogenized, the mineral isochron indicates the time of last closure of the minerals, but the age of the rocks cannot be determined by the Rb-Sr method.

5. Chemical Weathering

All of the above conclusions regarding the suitability for dating of rocks and minerals apply only when the rocks or their minerals have not been altered by chemical weathering at or near the surface of the Earth. Because most rocks that are used for dating are usually collected from outcrops, the effects of chemical weathering on the ^{87}Rb-^{87}Sr decay scheme may be important.

The effect of chemical weathering on the Rb-Sr decay systematics in rocks and minerals has been studied by GOLDICH and GAST (1966), BOTTINO and FULLAGAR (1968), and DASCH (1969). Understanding of this process is important to an evaluation of the reliability of dates obtained by the Rb-Sr method and to the geochemistry of strontium isotopes generally.

DASCH (1969) measured the concentrations of rubidium and strontium and $^{87}Sr/^{86}Sr$ ratios in several weathering profiles developed on igneous and metamorphic rocks. He found that progressive chemical weathering caused an increase in the Rb/Sr ratio due to loss of strontium and concurrent increase of the rubidium concentration of the rocks. The $^{87}Sr/^{86}Sr$ ratio, however, was found to remain constant, or increased only slightly. BOTTINO and FULLAGAR (1968) reported similar results for the Petersburg Granite of

Greenville County, Virginia. They showed that in the weathered zone lithium and calcium are depleted by about 50 percent, while sodium, magnesium, and strontium are depleted by about 25 percent compared to fresh rock samples. Potassium showed no change, but rubidium was enriched by about 20 percent. The average Rb/Sr ratio of weathered rock was found to be about 70 percent higher than that of fresh rock. Chemical weathering of igneous and metamorphic rocks therefore tends to increase the Rb/Sr ratios, moving them to the right on an isochron diagram. As a result, dates derived from chemically weathered rocks or minerals by the Rb-Sr method may be lowered appreciably.

A spectacular example of the effect of chemical weathering on Rb-Sr and K-Ar dates of biotite was reported by GOLDICH and GAST (1966). They separated biotite from residual clay formed by chemical weathering of the underlying Morton Gneiss near North Redwood, Minnesota. Rb-Sr dates of these biotites are about 75 percent lower than dates derived from un-weathered biotite samples. The K-Ar dates of the weathered biotites, however, are only 25 percent lower, which presents an interesting anomaly.

The lowering of Rb-Sr dates of biotites compared to K-Ar dates has been attributed to a process of base exchange by KULP and ENGELS (1963) and KULP (1967). These authors suggested that calcium-bearing ground-water can remove potassium and radiogenic ^{40}Ar quantitatively from successive cleavage flakes of biotite, leaving the $^{40}Ar/^{40}K$ ratio of the rest of the mineral unchanged. In this process radiogenic ^{87}Sr may be exchanged for other strontium isotopes in the water and is removed from the mineral. In addition, Rb^+ ions may replace K^+ ions, thereby increasing the rubidium concentration of the biotite. The result of these exchange reactions would be that the Rb-Sr dates of biotite are lowered significantly, while the K-Ar dates remain unaffected. Experiments with biotite under laboratory conditions suggested that 50 percent of the potassium can be removed together with associated radiogenic ^{40}Ar without affecting K-Ar dates. Rb-Sr dates, on the other hand, were lowered significantly by treating biotite with a solution containing rubidium and calcium. KULP and ENGEL (1963, p. 234) concluded from these experiments that, "If sufficient rubidium is present in groundwater, rubidium will be added to the mica while radiogenic ^{87}Sr is removed so that the Rb-Sr age will be lowered relative to the unchanged K-Ar age".

BROOKS (1968) has investigated the open-system behavior of co-existing plagioclase and potassium-feldspar from the Heemskirk granite of western Tasmania (see Chapter V). He was able to show that altered plagioclase had gained significant amounts of radiogenic ^{87}Sr, ^{86}Sr, and ^{87}Rb which had been derived from the potassium-feldspar. The significance of these results is that the alteration involved not only the movement of radiogenic ^{87}Sr, but also of rubidium and normal strontium.

6. Summary

Metamorphism of crystalline rocks disturbs the Rb-Sr decay systematics of their minerals. Whole-rock samples of normal hand specimen size may remain closed or may also be affected. The alteration resulting from metamorphism may involve migration of radiogenic ^{87}Sr as well as of rubidium and strontium. Useful information about the dates of original crystallization and subsequent metamorphism can be derived by means of separate whole-rock and mineral isochrons, provided the whole-rock samples remained closed and that the minerals had reached isotopic equilibrium prior to closure. The study of systematic variations of mineral dates in contact metamorphic aureoles is fruitful because it permits a more quantitative treatment of the physical processes which play a role in resetting the mineral clocks. Chemical weathering of granitic igneous rocks tends to deplete them in strontium and enrich them in rubidium, causing a lowering of dates calculated from such rocks. Biotite is very susceptible in this regard, perhaps because of base-exchange reactions with groundwater which cause loss of radiogenic ^{87}Sr and an increase of the rubidium concentration. It is well to remember that rocks to be dated by the Rb-Sr method rarely have the simple histories we wish them to have and that they may approach, but rarely satisfy, the assumption that they have remained closed systems from the time they formed to the time they are taken to the laboratory for analysis.

X. Meteorites

1. Introduction

Meteorites are solid objects that have fallen to the Earth from space. They consist primarily of silicate minerals, or iron-nickel alloys, or both, and they are classified accordingly as stones, irons, or stony irons. The stony meteorites are made up of olivine, orthopyroxene, and small amounts of other silicate minerals such as feldspar and diopside. The irons are composed mainly of the iron-nickel alloys taenite and kamacite, but they also contain troilite (FeS) and inclusions of silicate minerals. The stony-irons consist of roughly equal proportions of metallic and silicate phases.

The stony meteorites are divided into two groups: chondrites and achondrites. The *chondrites* are so named because they contain chondrules — spheroids about 1 millimeter in diameter composed of silicate minerals with some glass and metal. The origin of chondrules has been the subject of discussion for more than a century. They were first examined in thin section by SORBY (1877), who concluded that they were devitrified droplets of glass which had cooled rapidly as a "fiery rain". WHIPPLE (1966) suggested that the chondrules formed in the cooler parts of the solar nebula as the result of fusion of dust by lightning. Whatever their origin, it is very likely that the chondrules predate the stony meteorites in which they are preserved, and therefore they may be among the oldest physical objects in the solar system. Phase relations of the silicate minerals of chondrules indicate that they were later heated and either partially or completely recrystallized.

The chondrites are subdivided into three main classes on the basis of chemical and mineralogic criteria. These are: (1) enstatite chondrites, (2) ordinary chondrites, and (3) carbonaceous chondrites. The ordinary chondrites are further subdivided on the basis of compositional differences into the bronzite chondrites, the hypersthene chondrites, and the amphoterites. VAN SCHMUS and WOOD (1967) have devised an even more detailed classification which recognizes six petrologic types for each of the five chemical classes, or 30 possible varieties of chondritic meteorites.

The *achondrites* do not contain chondrules and have many features in common with terrestrial igneous rocks. They are subdivided into calcium-rich and calcium-poor groups. Their mineralogy and texture indicate that they crystallized from a silicate melt under conditions which permitted segregation of feldspar from olivine and pyroxene. They appear to be the most highly-modified group of stony meteorites.

Several hypotheses have been proposed to account for the origin of meteorites. They are based on (1) a single parent body of planetary size; (2) two successive generations of parent bodies, one of lunar size and one of asteroidal size; or (3) many bodies of asteroidal size. The single-body hypothesis, which has been advocated by RINGWOOD (1959) and LOVERING (1957, 1958), among others, appeals to geologists because it readily allows analogies between the meteorite parent body and the Earth. For example, DALY (1943) attempted to deduce the evolution and internal structure of the Earth by an analogy with a meteorite parent body of planetary size. However, the single-parent body hypothesis meets with several serious objections (ANDERS and GOLES, 1961). The hypothesis of two successive generations of parent bodies has been developed primarily by UREY (1956, 1957, 1958). It succeeds in explaining most of the properties of meteorites, but relies on events which seem to have low probabilities of occurrence. The third hypothesis suggests that the meteorites are fragments resulting from collisions among the asteroids and that the asteroids are, in fact, the parent bodies in which the meteorites acquired their mineral compositions and structures (FISH, GOLES, and ANDERS, 1960).

MASON (1967) has summarized the history of the formation of meteorites as a series of eight distinct events: "(1) nucleosynthesis; (2) formation of the ancestral solar nebula, with a central proto-sun surrounded by a lens-shaped cloud of dust and gas, diminishing in temperature from the center to the margin; (3) formation of chondrules; (4) aggregation of chondrules and dust into asteroidal-sized bodies; (5) heating-up of these bodies to varying degrees, depending largely on their size, with partial or complete melting of some; (6) cooling and crystallization of molten material, with differentiation of nickel-iron and silicates; (7) breakup of meteorite parent bodies; (8) arrival of meteorites on the Earth."

More information on meteorites is available in books by WOOD (1968), MASON (1962), and KRINOV (1960), and in the review papers of MASON (1967), RINGWOOD (1966), ANDERS (1964), WOOD (1963), and ANDERS and GOLES (1961). In the following sections of this chapter we shall discuss the results of measurement of the isotopic composition of strontium in stony and iron meteorites.

2. Rubidium and Strontium in Stony Meteorites

More than 70 stony meteorites have been analyzed for rubidium and strontium by isotope dilution. In addition, several different fragments of some meteorites have been analyzed, raising the total number of such analyses to over 115. This count does not include analyses of minerals separated from stony or iron meteorites. Nevertheless, the figures are impressive and reflect the intense efforts which have been made to date

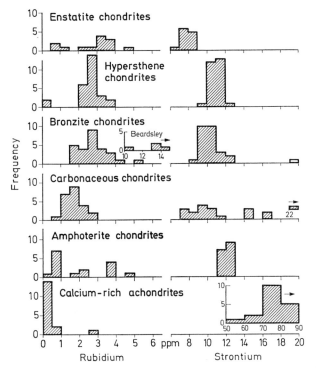

Fig. X.1. Rubidium and strontium concentrations of chrondrites and calcium-rich achondrites. All of the meteorites are falls. The classes of chondrites are arranged in order of decreasing rubidium content. Where more than one specimen of a meteorite has been analyzed, all have been included. However, the diagram does not include analyses of mineral concentrates or of specific gravity fractions. The data were taken from the literature referred to in the text

meteorites by the Rb-Sr method. The distribution of rubidium and strontium in stony meteorites (falls only) is displayed in Fig. X.1.

The first observation to be made from Fig. X.1 is that the chondrites and the calcium-rich achondrites have very different concentrations of rubidium and strontium. Most chondrites contain between 9 and 12 ppm strontium and between 1 and 4 ppm rubidium. The calcium-rich achrondrites, on the other hand, have strontium contents between 70 and 85 ppm, but usually contain less than 1 ppm rubidium. In general, the rubidium concentration within each group tends to be more variable than the strontium concentration. Only three calcium-poor achondrites (Johnstown, Bishopville, and Norton County) have been analyzed for rubidium and strontium. These three meteorites are unusually low in strontium (1 to 15 ppm), and also rather low in rubidium (less than 2 ppm).

Several stony meteorites show remarkable inhomogeneity. The amphoterite chondrite Soko-Banja, a "polymict breccia" whose fall in Yugoslavia was reported in 1879, is a good example. GOPALAN and WETHERILL (1969) analyzed three fragments of this meteorite and found that its rubidium content ranges from 0.5 to 4.88 ppm, whereas strontium varies only from 10.48 to 11.23 ppm. Nevertheless, the three fragments which have been analyzed appear to have a common age of 4.5 billion years, suggesting that Soko-Banja acquired its heterogeneity a long time ago.

When meteorites are exposed to chemical weathering on the surface of the Earth, their rubidium and strontium concentrations may be significantly affected. This was first demonstrated experimentally by GAST (1962), who showed that substantial fractions of rubidium and strontium could be leached with water and dilute acetic acid from the bronzite chondrite Beardsley, which fell on October 15, 1929, in Rawlins County, Kansas. Though some fragments were collected on the day of the fall and are therefore unweathered, others (such as Beadsley I of GAST) were collected more than a year later and appear to have been chemically altered. Another example of alteration was observed in the amphoterite chondrites Lake Labyrinth, Oberlin, and Chico, all of which are finds. None of them fits the isochron reported by GOPALAN and WETHERILL (1969), in spite of the fact that the specimens looked fresh and were taken from the interiors of the

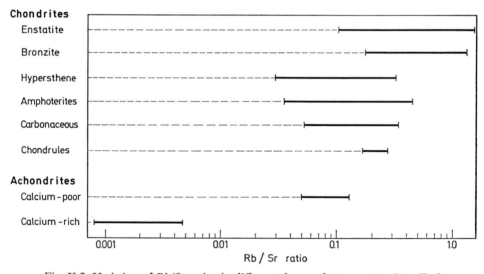

Fig. X.2. Variation of Rb/Sr ratios in different classes of stony meteorites. Each of the five classes of chondrites has been dated separately by the Rb/Sr method. The calcium-rich achondrites have by far the lowest Rb/Sr ratios of any class of stony meteorites. The data for this diagram were taken from the literature referred to in the text

available fragments. At least two of these (Lake Labyrinth and Oberlin) are anomalously enriched in strontium, and Chico may have been depleted in rubidium. A third case of terrestrial contamination of a meteorite was reported by SHIELDS, PINSON, and HURLEY (1966) for the chondrite Bjurböle, which fell through sea ice in Finland on March 12, 1899. Although several fragments were soon recovered, this meteorite later proved to have been hopelessly contaminated.

We showed in Chapter II that one of several prerequisites of the Rb-Sr isochron method of dating is that the objects to be dated must exhibit a sufficiently-wide variation in Rb/Sr (or $^{87}Rb/^{86}Sr$) ratios. The wider the range of Rb/Sr ratios, the more precisely the slope of the isochron can be determined. Fig. X.2 shows the Rb/Sr ratios of the different classes of stony meteorites. The chondrites have Rb/Sr ratios ranging from 0.03 to 1.5, while those of the calcium-rich achondrites range from 0.0008 to 0.0047. Fortunately, therefore, the Rb/Sr ratios of the chondrites and achondrites vary sufficiently to permit them to be dated by the Rb-Sr isochron method. In the following section we shall review the results of Rb-Sr age determinations of meteorites.

3. The Ages of Stony and Iron Meteorites

The problem of dating stony meteorites by the Rb-Sr method can be approached in three different ways. First, one can test the hypothesis that all types of stony meteorites evolved in a relatively-short interval of time from the same mass of isotopically homogeneous primordial material (Model 1). If so, chondrites and achondrites should fit a single isochron whose slope and intercept yield their apparent age and initial $^{87}Sr/^{86}Sr$ ratio, as discussed in Chapter II. Second, it may be true that all meteorites of a particular class had a common origin, but that different classes of meteorites formed at different times from primordial material having different $^{87}Sr/^{86}Sr$ ratios (Model 2). Third, each meteorite may have a unique age and initial $^{87}Sr/^{86}Sr$ ratio which can be determined by analysis of its constituent minerals, provided that the strontium in all of its phases was isotopically homogeneous at some time in its past (Model 3). Differences in the apparent ages and initial $^{87}Sr/^{86}Sr$ ratios of individual meteorites may provide information about possible postcrystallization events, and thus about the early history of the solar system.

Meteorites have also been dated by other methods which will not be discussed here. The results of these studies have been reviewed by ANDERS (1962, 1963), KRANKOWSKY and ZÄHRINGER (1966), CUMMING (1969), and BURNETT (1971).

The different assumptions required to date meteorites by the Rb-Sr method also require increasingly precise analyses. For this reason, the efforts to date meteorites have progressed historically from initial attempts at dating stony meteorites according to the assumptions of Model 1 to the more refined treatment implied by Model 3. The review of this work will follow this historical path in order to emphasize the considerable achievements which have been made and to suggest the direction in which future progress is likely to be made.

The first age determinations of stony meteorites by the Rb-Sr method were made using Model 1 by combining analyses of chondrites and calcium-rich achondrites. Among the first to publish his results was SCHUMACHER (1956), who analyzed the chondrite Forest City and the achondrites Pasamonte and Bustee. By assuming that these meteorites had formed at nearly the same time from the same primordial material, SCHUMACHER calculated a common age of $(4.8 \pm 0.4) \times 10^9$ years. HERZOG (1955) and HERZOG and PINSON (1956) obtained a similar date for the chondrites Homestead and Forest City and the achondrite Pasamonte. These results were confirmed in the following years by WEBSTER, MORGAN, and SMALES (1957), who reported a date of $(4.6 \pm 0.5) \times 10^9$ years for Forest City and the achondrite Johnstown. The most extensive study of this type was made by GAST (1962). Based on analyses of four calcium-rich achondrites and five bronzite and hypersthene chondrites, he obtained a date of 4.67×10^9 years and an initial $^{87}Sr/^{86}Sr$ ratio of 0.6975 (normalized to $^{86}Sr/^{88}Sr = 0.1194$). Later PINSON et al. (1965) reported a date of $(4.52 \pm 0.12) \times 10^9$ years and an initial ratio of 0.698 ± 0.001 for another suite of chondrites and achondrites. Their isochron for five chondrites and two calcium-rich achondrites is shown in Fig. X.3.

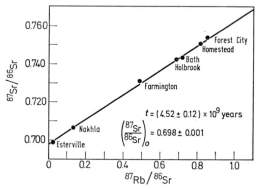

Fig. X.3. Rubidium-strontium isochron of stony meteorites, based on the assumption that the chondrites and achondrites formed in a relatively-short interval of time from primordial material whose strontium was isotopically homogeneous. The apparent fit of the points on the isochron supports these assumptions, but does not exclude the possibility that small unresolved differences in the ages and initial $^{87}Sr/^{86}Sr$ ratios of different classes of stony meteorites may exist. (Data from PINSON et al., 1965)

These results suggested that Model 1 is at least approximately correct and that stony meteorites were formed in a relatively short interval of time about 4.6×10^9 years ago from primordial matter that had a uniform $^{87}Sr/^{86}Sr$ ratio of approximately 0.698 to 0.700. However, GAST (1962) also considered the possibility that the achondrites may have formed from the chondrites, and he calculated an age for the chondrites alone.

The effort to date chondrites of a particular class without relying on assumed genetic relationships with other stony meteorites (Model 2) was continued by MURTHY and COMPSTON (1965). They analyzed four carbonaceous chondrites and obtained a date of $(4.46 \pm 0.35) \times 10^9$ years and an initial $^{87}Sr/^{86}Sr$ ratio of 0.7007. More recently, KAUSHAL and WETHERILL (1970) reported that five of a total of nine carbonaceous chondrites have a common age of 4.69×10^9 years and an initial ratio of 0.6983, consistent with bronzite chondrites. The other four meteorites do not fall on the bronzite isochron, possibly because of alteration of their rubidium and strontium content. The remaining classes of chondrites were dated by GOPALAN and WETHERILL (1968, 1969), KAUSHAL and WETHERILL (1969), and GOPALAN and WETHERILL (1970). PAPANASTASSIOU and WASSERBURG (1969), in what must be regarded as a breakthrough in Rb-Sr geochronometry, succeeded in dating a suite of calcium-rich achondrites whose total range of $^{87}Sr/^{87}Sr$ ratio was only 0.2 percent — from about 0.6993 to about 0.7005. Their isochron is shown in Fig. X.4. The instrumental technique used to gain such high precision was described by WASSERBURG et al. (1969).

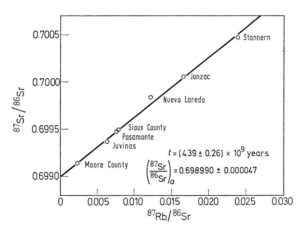

Fig. X.4. Rubidium-strontium isochron for calcium-rich achondrites. The common age of these achondrites is $(4.39 \pm 0.26) \times 10^9$ years, and their initial $^{87}Sr/^{86}Sr$ ratio is 0.698990 ± 0.000047. Dating of high-calcium achondrites was made possible by the use of a programmable mass spectrometer with digital data output connected directly to a computer. (Data taken from PAPANASTASSIOU and WASSERBURG, 1969)

The age determinations and initial ^{87}Sr/^{86}Sr ratios of different classes of chondrites and achondrites are compiled in Table X.1. The dates range from $(4.39 \pm 0.26) \times 10^9$ years for calcium-rich achondrites to $(4.69 \pm 0.14) \times 10^9$ years for bronzite chondrites. These results suggest that the different classes of meteorites formed within an interval of time from 4.5 to 4.7 billion years ago, which is not yet resolvable by the Rb-Sr isochron method of dating whole meteorites.

The initial ^{87}Sr/^{86}Sr ratios determined from the isochrons presumably record the isotopic composition of strontium of the parent bodies at the time of their final cooling. The most precisely-known initial ^{87}Sr/^{86}Sr ratio is that of the achondrites analyzed by PAPANASTASSIOU and WASSERBURG (1969). Their value for this ratio is 0.698990 ± 0.000047 and is referred to as BABI (basaltic achondrite best initial). The initial ratios of the other classes of stony meteorites have much larger errors. The results suggest that the strontium in the parent bodies was isotopically homogeneous within fairly narrow limits. However, the data in Table X.1 do not exclude the possibility that different classes of meteorites or individual stones had small, but significantly different, initial ^{87}Sr/^{86}Sr ratios.

Considerably more detail about the histories of meteorites can be derived from internal isochrons based on separated mineral fractions of individual stones (Model 3). The first such age determination was attempted by

Table X.1. Rubidium-strontium isochron ages and initial ^{87}Sr/^{86}Sr ratios of different classes of chondrites and achondrites (Model 2)

Class	Age in billions of years	Initial ^{87}Sr/^{86}Sr		Reference
Carbonaceous chondrites	4.46 ± 0.35	0.7007		MURTHY and COMPSTON (1965)
Carbonaceous chondrites	4.69[a]	0.6983[a]		KAUSHAL and WETHERILL (1970)
Hypersthene chondrites	4.48 ± 0.14	0.7008	± 0.0012	GOPALAN and WETHERILL (1968)
Hypersthene chondrites	4.54 ± 0.12	0.7003	± 0.0004	KAUSHAL and WETHERILL (1970)
Enstatite chondrites	4.54 ± 0.13	0.6993	± 0.0015	GOPALAN and WETHERILL (1970)
Amphoterite chondrites	4.56 ± 0.15	0.7005	± 0.0015	GOPALAN and WETHERILL (1969)
Bronzite chondrites	4.69 ± 0.14	0.6983	± 0.0024	KAUSHAL and WETHERILL (1969)
Calcium-rich achondrites	4.39 ± 0.26	0.698990 ± 0.000047		PAPANASTASSIOU and WASSERBURG (1969)

[a] Consistent with bronzite isochron of KAUSHALL and WETHERILL (1969).
All dates are based on $\lambda (^{87}$Rb$) = 1.39 \times 10^{-11}$ yr^{-1}.

Compston, Lovering, and Vernon (1965) for the calcium-poor enstatite achondrite Bishopville. They analyzed the total meteorite and separated enstatite and feldspar (oligoclase) fractions and obtained an isochron, shown in Fig. X.5, consistent with a date of $(3.7 \pm 0.2) \times 10^9$ years and an initial $^{87}Sr/^{86}Sr$ ratio of 0.7030 ± 0.0020 (normalized to $^{86}Sr/^{88}Sr = 0.1194$). This

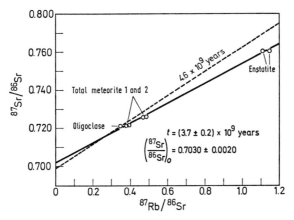

Fig. X.5. Internal Rb-Sr isochron for the low-calcium enstatite achondrite Bishopville. The apparent age of this meteorite, $(3.7 \pm 0.2) \times 10^9$ years, is lower than the ages obtained from suites of meteorite whole-rock samples by Models 1 and 2. This date can be interpreted either as the time of the initial crystallization of Bishopville or as the time of metamorphic homogenization of the strontium among its minerals. Unfortunately, the reliability of the date is lessened by its dependence on the analysis of the enstatite. (Data from Compston, Lovering, and Vernon, 1965)

age determination is based only on the assumptions that (1) the $^{87}Sr/^{86}Sr$ ratios of the minerals and the total meteorite were the same at some time in the past, and (2) the minerals have remained closed to rubidium and strontium from that time until the present.

The apparent age of Bishopville is lower than the Model 1 or Model 2 ages of stony meteorites discussed above, and it is not entirely clear how this date should be interpreted. It can be seen in Fig. X.5 that the slope of the Bishopville isochron is strongly dependent on the two enstatite points. Burnett and Wasserburg (1967b) have questioned the reliability of these enstatite analyses. Some caution in interpreting the internal age of Bishopville may thus be in order until the mineral analyses are corroborated by additional work.

A similar attempt to date an individual meteorite was made by Shields, Pinson, and Hurley (1966). They analyzed separated phases of the chondrite Bjurböle, but found that this meteorite may have been altered by seawater after its fall. However, chondrules from Bjurböle gave dates that are consistent with the Model 1 ages of stony meteorites. An earlier attempt by MURTHY and COMPSTON (1965) to date the chondrules of Peace River failed.

Subsequent efforts to date individual meteorites by the Rb-Sr method have proceeded along two different lines: SHIMA and HONDA (1967) obtained internal isochrons for three stony meteorites, using a solvent-extraction technique to leach rubidium and strontium. Other workers successfully analyzed separated mineral fractions of meteorites — for example the enstatite achondrite Norton County (BOGARD et al., 1967) and the amphoterite chondrite Krähenberg (KEMPE and MÜLLER, 1969). In fact, Krähenberg is among the most precisely-dated meteorites. It crystallized (4.70 ± 0.014) × 10^9 years ago with an initial $^{87}Sr/^{86}Sr$ ratio of 0.6989 ± 0.0005. The isochron for Krähenberg is shown in Fig. X.6. The results of age determinations of individual meteorites are compiled in Table X.2.

WASSERBURG, BURNETT, and FRONDEL (1965) made the important discovery that the minerals making up the silicate inclusions in certain iron meteorites had sufficiently different Rb/Sr ratios to make them suitable for dating. In a subsequent paper BURNETT and WASSERBURG (1967a) analyzed silicate minerals from the iron meteorite Kodaikanal, one of the few meteorites in which potassium feldspar has been found. Feldspar glass from this meteorite contains 1260 ppm rubidium and 23.9 ppm strontium, giving it a Rb/Sr ratio of about 52.7. This is by far the highest Rb/Sr ratio ever

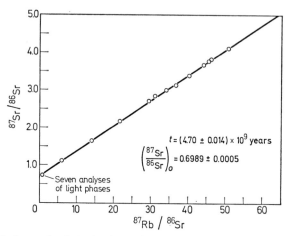

Fig. X.6. Internal Rb-Sr isochron for the amphoterite chondrite Krähenberg, based on analyses of its light and dark phases. The excellent precision of the date (t = (4.70 ± 0.014) × 10^9 years) indicated by this isochron results from the statistical analysis of separated fractions of the dark phases of this meteorite which have unusally high $^{87}Rb/^{86}Sr$ ratios. (Data are taken from KEMPE and MÜLLER, 1969)

Table X.2. Internal rubidium/strontium ages (Model 3) of individual stony and iron meteorites

Name	Class	Age in billions of years	Initial $^{87}Sr/^{86}Sr$	Reference
		Stony meteorites		
Bishopville	Enstatite achondrite	3.7 ± 0.2	0.7030 ± 0.0020	Compston, Lovering, and Vernon (1965)
Peace River	Hypersthene	4.5	0.697	Shima and Honda (1967)
Bruderheim	Hypersthene	4.54	0.6985	Shima and Honda (1967)
Abee	Enstatite	4.52	0.700	Shima and Honda (1967)
Norton County	Enstatite achondrite	4.70 ± 0.10	0.700 ± 0.002	Bogard et al. (1967)
Krähenberg	Amphoterite	4.70 ± 0.014	0.6989 ± 0.0005	Kempe and Müller (1969)
Soko Banja	Amphoterite	4.50	0.7005	Gopalan and Wetherill (1969)
Olivenza	Amphoterite	4.63 ± 0.16	0.6994 ± 0.0017	Sanz and Wasserburg (1969)
Indarch	Enstatite	4.56 ± 0.08	0.7005 ± 0.003	Gopalan and Wetherill (1970)
Guareña	Bronzite chondrite	4.56 ± 0.08	0.69995 ± 0.00015	Wasserburg, Papanastassiou, and Sanz (1969)
		Iron meteorites		
Kodaikanal	Octahedrite	3.8 ± 0.1	0.713 ± 0.0020	Burnett and Wasserburg (1967a)
Toluca	Octahedrite	4.7 ± 0.5	0.698 ± 0.002	Burnett and Wasserburg (1967b)
Weekeroo Station	Octahedrite	$4.37 \begin{array}{l}+0.23\\-0.12\end{array}$	$0.703 \begin{array}{l}+0.002\\-0.003\end{array}$	Burnett and Wasserburg (1967b)
Odessa	Octahedrite	4.5	0.699 (assumed)	Burnett and Wasserburg (1967b)
Linwood	Octahedrite	4.55	0.699 (assumed)	Burnett and Wasserburg (1967b)
Pine River	Octahedrite	4.55	0.699 (assumed)	Burnett and Wasserburg (1967b)
Four Corners	Octahedrite	4.6	not determined	Burnett and Wasserburg (1967b)
Colomera	Octahedrite	4.6	not determined	Burnett and Wasserburg (1967b)
Colomera	Octahedrite	4.61 ± 0.04	0.69940 ± 0.00004	Sanz, Burnett, and Wasserburg (1970)

All dates are based on $\lambda \, (^{87}Rb) = 1.39 \times 10^{-11} \, yr^{-1}$.

found in meteoritic material. As a consequence, Burnett and Wasserburg were able to determine a very precise date of 3.69×10^9 years and an initial $^{87}Sr/^{86}Sr$ ratio of 0.758 for Kodaikanal. However, the feldspar glass point appears to lie significantly below the isochron, shown in Fig. X.7, formed by the less rubidium-enriched fractions. The best estimate of the age of Kodaikanal is obtained when the feldspar-glass point is ignored. The

resulting date is $(3.8 \pm 0.1) \times 10^9$ years, and the initial ratio is 0.713 ± 0.020
The date indicated by Kodaikanal was interpreted by BURNETT and WAS-
SERBURG (1967a) as the time of final differentiation and cooling of its parent
body, which seems to have remained hot long after the others had cooled
and become inactive.

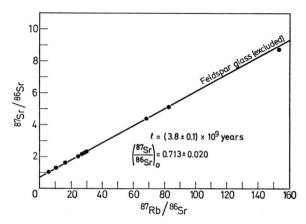

Fig. X.7. Rubidium-strontium isochron for the silicate minerals of the iron meteorite Kodaikanal. This meteorite gives a date of $(3.8 \pm 0.1) \times 10^9$ years and
therefore appears to be significantly younger than any other meteorite, with the
possible exception of Bishopville, and Weekeroo Station. (Data are taken from
BURNETT and WASSERBURG, 1967a)

4. The Initial $^{87}Sr/^{86}Sr$ Ratios of Stony Meteorites

The initial $^{87}Sr/^{86}Sr$ ratios of stony meteorites are important for two
reasons. First, the planet Earth and the parent bodies of meteorites are
believed to have been formed at about the same time by gravitational accretion of the condensed phases of the solar nebula. It is reasonable to assume
as a first approximation that the strontium which was incorporated into the
planet Earth had the same isotopic composition as that incorporated into
the parent bodies of the meteorites. The $^{87}Sr/^{86}Sr$ ratio of this primordial
strontium can therefore be regarded as the starting point for the isotopic
evolution of terrestrial and meteoritic strontium. The value of this primordial ratio can be determined only by analysis of strontium-bearing rocks
or minerals that crystallized at or very soon after the time of initial formation
of the planets. Because of the continuing geological activity and chemical
differentiation of the planet Earth, it is unlikely that any terrestrial rocks or
minerals that formed at the very beginning will have survived, and none
have yet been found. On the other hand, most of the parent bodies of the

stony meteorites appear to have become chemically-inactive a relatively short time after their formation, presumably because their small size prevented them from accumulating and retaining heat. As a consequence, stony meteorites, taken as a group, can be used to determine the value of the primordial ^{87}Sr/^{86}Sr ratio for both terrestrial and meteoritic strontium.

The second reason for our interest in the initial ^{87}Sr/^{86}Sr ratios of meteorites is that they provide evidence of the sequence of significant events of short duration during the early history of meteorites and their parent bodies. PAPANASTASSIOU and WASSERBURG (1969) and WASSERBURG, PAPANASTASSIOU, and SANZ (1969) have provided the theoretical basis for the interpretation of their very precisely-determined initial ^{87}Sr/^{86}Sr ratios of the achondrites and of individual meteorites, such as the chondrite Guareña.

Based on the available data, it is reasonable to conclude that planetary objects, including the parent bodies of the meteorites, formed from the solar nebula during a short interval of time. The ^{87}Sr/^{86}Sr ratio of the solar nebula was increasing as a result of decay of ^{87}Rb to ^{87}Sr at a rate proportional to its Rb/Sr ratio. At some time in the past, its ^{87}Sr/^{86}Sr ratio was equal to BABI — which is the most precisely-determined and also among the lowest ^{87}Sr/^{86}Sr ratios ever measured in meteorites. BABI thus provides a reference point relative to which the isotopic evolution of strontium of different objects in the solar system can be evaluated. PAPANASTASSIOU (1970) obtained a slightly lower initial ^{87}Sr/^{86}Sr ratio of 0.69884 $\pm$ 0.00004 (ADOR) for the augite achondrite Angra dos Reis (Rio de Janeiro, January 30, 1869).

In the case of Guareña, WASSERBURG, PAPANASTASSIOU, and SANZ (1969) obtained an internal isochron date of 4.56 $\pm$ 0.08 billion years and an initial ^{87}Sr/^{86}Sr ratio of 0.69995 $\pm$ 0.00015. The initial ^{87}Sr/^{86}Sr ratio of Guareña is significantly greater than BABI, which indicates that this meteorite may have been metamorphosed sometime after its formation. The question is: How long did it take the strontium in Guareña to evolve from BABI to its observed initial ^{87}Sr/^{86}Sr ratio? The answer to this question can be obtained by the calculation outlined below.

We showed in Chapter II [Eq. (II.15)] that to a good approximation:

$$\frac{^{87}\text{Sr}}{^{86}\text{Sr}} \cong \left(\frac{^{87}\text{Sr}}{^{86}\text{Sr}}\right)_0 + \frac{^{87}\text{Rb}}{^{86}\text{Sr}}\,\lambda t. \tag{X.1}$$

In order to adapt this equation to this particular case, we write

$$\left(\frac{^{87}\text{Sr}}{^{86}\text{Sr}}\right)_i = \text{BABI} + \left(\frac{^{87}\text{Rb}}{^{86}\text{Sr}}\right)_i \lambda \Delta t, \tag{X.2}$$

where

$\left(\dfrac{^{87}\text{Sr}}{^{86}\text{Sr}}\right)_i$ = the initial ^{87}Sr/^{86}Sr ratio of the meteorite as measured by its internal isochron;

$\left(\dfrac{^{87}\text{Rb}}{^{86}\text{Sr}}\right)_i =$ the ratio of these isotopes in the whole meteorite at the time of last isotopic equilibration indicated by its internal isochron;

$\varDelta t =$ the time interval during which the strontium in the meteorite evolved from BABI to $(^{87}\text{Sr}/^{86}\text{Sr})_i$.

This calculation, when applied to Guarena, results in $\varDelta t = 74 \pm 12$ million years, which illustrates the important point that precisely-determined initial $^{87}\text{Sr}/^{86}\text{Sr}$ ratios can be used to detect the occurrence of events of short duration very early in the history of meteorites.

Similar reasoning was used by PAPANASTASSIOU and WASSERBURG (1969) to set limits on the time interval within which achondrites may have separated from the solar nebula. According to LAMBERT and WARNER (1968) and LAMBERT and MALLIA (1968) the Rb/Sr ratio of the photosphere of the sun is 0.65 ± 0.33. Assuming that this value is representative of the Rb/Sr ratio of the solar nebula, one may use Eq. (II.16) to calculate the increase of its $^{87}\text{Sr}/^{86}\text{Sr}$ ratio as a function of time:

$$\frac{^{87}\text{Sr}}{^{86}\text{Sr}} - \left(\frac{^{87}\text{Sr}}{^{86}\text{Sr}}\right)_0 \simeq k\left(\frac{\text{Rb}}{\text{Sr}}\right)\lambda t. \qquad (\text{X.3})$$

Fig. X.8 is a plot of this equation for systems having Rb/Sr ratios of 0.65 and 0.25, which is the value of this ratio for chondritic meteorites.

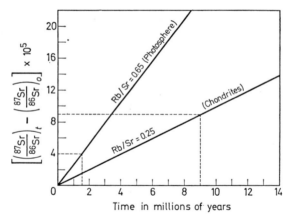

Fig. X.8. The resolution of the time interval during which the parent bodies of calcium-rich achondrites separated from the solar nebula. According to PAPANA-STASSIOU and WASSERBURG (1969), the maximum difference among the initial $^{87}\text{Sr}/^{86}\text{Sr}$ ratios of their suite of calcium-rich achondrites is 9×10^{-5} (Fig. X.4). The corresponding interval of time during which the parent bodies of these meteorites separated from the solar nebula depends on the Rb/Sr ratio of the nebula, as shown in the diagram. An upper limit of 9 million years is obtained by setting the Rb/Sr ratio of the solar nebula equal to 0.25 (chondrites) and by taking the maximum difference among the initial $^{87}\text{Sr}/^{86}\text{Sr}$ ratios of achondrites equal to 9×10^{-5}. The lower limit of 1.6 million years is based on a Rb/Sr ratio of 0.65 (photosphere of the sun) and by taking the smallest difference in initial $^{87}\text{Sr}/^{86}\text{Sr}$ ratios

The precision of measured $^{87}Sr/^{86}Sr$ ratios achieved by PAPANASTASSIOU and WASSERBURG (1969) for calcium-rich achondrites is about 7×10^{-5}. The uncertainty in the initial $^{87}Sr/^{86}Sr$ ratio, however, depends also on the uncertainty in the age of the sample and may be worse than the precision of the measurement of the $^{87}Sr/^{86}Sr$ ratio. Using all of their data for calcium-rich achondrites, PAPANASTASSIOU and WASSERBURG (1969) concluded that the maximum possible difference in the initial ratios of their samples is 9×10^{-5}. Reference to Fig. X.8 shows that this difference corresponds to separation from the solar nebula within a time interval of about 9×10^{6} years for $Rb/Sr = 0.25$. The best time resolution was achieved by omitting Nuevo Laredo, which may have a slightly higher initial $^{87}Sr/^{86}Sr$ ratio than the other calcium-rich achondrites (see Fig. X.4). The maximum difference in the initial $^{87}Sr/^{86}Sr$ ratios of the remaining achondrites then reduces to 4×10^{-5}, corresponding to a period of only 1.6×10^{6} years for the separation of the calcium-rich achondrites from a solar nebula with a Rb/Sr ratio of 0.65. These calculations suggest, therefore, that the parent bodies of the achondrites separated from the solar nebula within only a few million years.

5. Summary

All available evidence suggests that stony and iron meteorites are fragments of larger parent bodies which formed during the earliest period of formation of the solar system. These parent bodies apparently cooled and became closed systems with regard to rubidium and strontium from 4.5 to 4.7 billion years ago. Only the iron meteorite Kodaikanal definitely formed more recently 3.8 ± 0.1 billion years ago.

The initial $^{87}Sr/^{86}Sr$ ratio of basaltic achondrites is 0.69899 ± 0.000047 and provides an important reference point for the isotopic evolution of strontium in the solar system. Small differences in the initial $^{87}Sr/^{86}Sr$ ratios of individual meteorites provide information regarding events of short duration during the very early stages of the history of meteorites.

XI. The Moon

1. Introduction

On July 20, 1969, NEIL A. ARMSTRONG and EDWIN E. ALDRIN landed successfully on the Moon in the Sea of Tranquillity at latitude 9.67° N and longitude 23.49° E. After exploring the lunar surface in the vicinity of their spacecraft they returned safely to Earth with 22 kilograms of rocks and dust they had collected. The exploration of the Moon and the collection of rock samples for analysis has been continued by subsequent American flights of the Apollo program and by the automatic probe "Luna 16" of the U.S.S.R.

One of the primary objectives of the Apollo program of the National Aeronautics and Space Administration of the U.S.A. is the dating of lunar rocks and minerals. Small portions of the lunar samples have been distributed to geochronologists in the U.S.A. and abroad for dating by the Rb-Sr, K-Ar, U-Pb and Th-Pb methods. Preliminary results for rocks from the Sea of Tranquillity were published in the so-called "Moon Issue" of *Science* (Vol.167 No. 3918, January 1970). More complete reports, including color photographs, were subsequently issued as a three-volume supplement to Volume 34 of *Geochimica et Cosmochimica Acta* in June of 1970.

The first age determinations of lunar rocks by means of the Rb-Sr method were reported by GOPALAN et al. (1970a); HURLEY and PINSON (1970a); COMPSTON et al. (1970a); MURTHY, SCHMITT, and REY (1970); WANLESS, LOVERIDGE, and STEVENS (1970a); GAST and HUBBARD (1970); and ALBEE et al. (1970). The last-named group from the California Institute of Technology is directed by G. J. WASSERBURG and adopted the *nom de plume* "Lunatic Asylum." A concise summary of the results of age determinations of lunar material has been published by WETHERILL (1971); our remarks are restricted to samples returned by the Apollo 11 and 12 missions.

2. Lunar Basalt

The rocks collected by the astronauts on the surface of the Moon are boulders and cobbles from the lunar regolith. No outcrops of bedrock have yet been sampled. Consequently, there is no justification for assuming *a priori* that the pieces of rock found in the lunar regolith at a particular location are cogenetic — that is, all have the same age and the same initial $^{87}Sr/^{86}Sr$ ratio. These rocks, therefore, must be dated individually by means of internal mineral isochrons.

Separated minerals as well as specific gravity and magnetic concentrates prepared from lunar basalts have been analyzed by ALBEE et al. (1970), COMPSTON et al. (1970a, b), PAPANASTASSIOU, WASSERBURG, and BURNETT (1970), Lunatic Asylum (1970), PAPANASTASSIOU and WASSERBURG (1970, 1971), and others. Table XI.1 contains a summary of the results by the Caltech group. According to PAPANASTASSIOU, WASSERBURG, and BURNETT (1970), internal isochrons for six basalts from the Sea of Tranquillity give dates in the range $(3.65 \pm 0.06) \times 10^9$ years, while COMPSTON et al. (1970a, b) reported dates of $(3.8 \pm 0.3) \times 10^9$ and $(3.80 \pm 0.11) \times 10^9$ years for mineral isochrons from two basalt specimens. The mineral isochrons are based on analyses not only of plagioclase and pyroxene, but also of cristobalite and ilmenite concentrates. In fact, the ilmenite fraction sometimes has the highest Rb/Sr ratio of any of the minerals in the lunar basalts. Fig. XI.1 is a mineral isochron for specimen 10044 from the Sea of Tranquillity.

Table XI.1. Internal rubidium-strontium isochron dates and initial $^{87}Sr/^{86}Sr$ ratios of basalt from the moon

Sample	Mineral isochron dates 10^9 years	Initial $^{87}Sr/^{86}Sr^\circ$	Reference
Sea of Tranquillity (Apollo 11)			
10071	3.68 ± 0.02	0.69926 ± 0.00003	PAPANASTASSIOU, WASSERBURG, and BURNETT (1970)
10057	3.63 ± 0.002	0.69939 ± 0.000004	PAPANASTASSIOU, WASSERBURG, and BURNETT (1970)
10017	3.59 ± 0.05	0.69932 ± 0.00005	PAPANASTASSIOU, WASSERBURG, and BURNETT (1970)
10069	3.68	0.69929	PAPANASTASSIOU, WASSERBURG, and BURNETT (1970)
10044	3.71 ± 0.11	0.69909 ± 0.00007	PAPANASTASSIOU, WASSERBURG, and BURNETT (1970)
10058	3.63 ± 0.20	0.69906 ± 0.00008	PAPANASTASSIOU, WASSERBURG, and BURNETT (1970)
10024	3.61 ± 0.07	0.69935 ± 0.00008	PAPANASTASSIOU and WASSERBURG (1971)
Ocean of Storms (Apollo 12)			
12002	3.36 ± 0.10	0.69949 ± 0.00005	PAPANASTASSIOU and WASSERBURG (1970)
12051	3.26 ± 0.10	0.69932 ± 0.00006	PAPANASTASSIOU and WASSERBURG (1970)
12064	3.18 ± 0.09	0.69943 ± 0.00006	PAPANASTASSIOU and WASSERBURG (1971)
12040	3.30 ± 0.04	0.69935 ± 0.00003	PAPANASTASSIOU and WASSERBURG (1971)
12035	3.20	0.69949	PAPANASTASSIOU and WASSERBURG (1971)
12036	3.30 ± 0.13	0.69918 ± 0.00006	PAPANASTASSIOU and WASSERBURG (1971)
12065	3.16 ± 0.09	0.69957 ± 0.00005	PAPANASTASSIOU and WASSERBURG (1971)
Basaltic achondrites (BABI)		0.69898 ± 0.00003	PAPANASTASSIOU (1970)
Angra dos Reis (ADOR)		0.69884 ± 0.00004	PAPANASTASSIOU (1970)

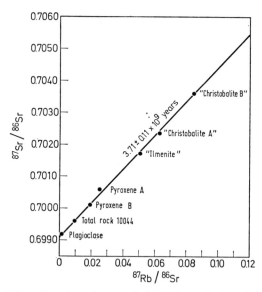

Fig. XI.1. Mineral isochron for sample 10044 from the Sea of Tranquillity. Mineral concentrates from this low-potassium basalt clearly define an isochron from which a date of 3.71 billion years can be calculated. This date indicates the time of crystallization of basalt in the Sea of Tranquillity. (Adapted from PAPANASTAS-SIOU, WASSERBURG, and BURNETT, 1970)

These dates refer to the last time when the strontium in the minerals was isotopically homogeneous and the minerals became closed to rubidium and strontium. They do not necessarily indicate the time elapsed since crystallization of the basalt, because it is possible that the mineral isochrons refer to a post-crystallization event which caused re-equilibration of the strontium isotopes in the minerals (see Chapter IX). However, the texture and chemical heterogeneity of the lunar basalts show no evidence of post-crystallization metamorphism. Moreover, the initial $^{87}Sr/^{86}Sr$ ratios of the basalts from Tranquillity Base fall into a narrow range from 0.6990 to 0.6994 and are equal to or higher than the basaltic achondrite best initial ratio (BABI $= 0.69898 \pm 0.00003$) as stated by PAPANASTASSIOU and WASSERBURG (1970). If the basalts had originally crystallized 4.6 billion years ago and had subsequently experienced closed-system metamorphism 3.6 billion years ago, the initial $^{87}Sr/^{86}Sr$ ratios of some of the basalts should be higher than the observed values or their original $^{87}Sr/^{86}Sr$ ratios must have been considerably less than BABI. Finally, several investigating teams [COMPSTON et al. (1970a, b), GOPALAN et al. (1970a, b), and GAST and HUBBARD (1970)] found that the basalts fit a whole-rock-isochron whose slope is very nearly identical to that of the mineral isochrons. On the basis of this evidence the date of 3.6 billion years is regarded as the time of crystallization of the basalts at Tranquillity Base.

The interpretation of whole-rock isochrons for basalts from the Sea of Tranquillity is complicated by the fact that these rocks tend to fall into two distinctive types on the basis of their chemical compositions. These differences were originally pointed out by COMPSTON et al. (1970a, b) who observed that specimens in Group I contained more Rb, K, Ba, Th, Ti, Na, P, S, Zr, Y, and rare earth elements than Group II, but had lower Al and Ca concentrations. Many other investigators have documented these differences. The important point here is that the basalt samples form two distinct clusters on the isochron diagram as though there were only two samples. This fact is shown in Fig. XI.2. The line joining the two clusters on the isochron diagram could represent mixtures of Groups I and II in differing proportions and would therefore have no real meaning in terms of time. However, because of the similarity of the mineral isochron dates and this whole-rock isochron date, the validity of the latter has been conceded.

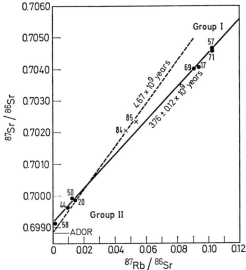

Fig. XI.2. Whole-rock isochron for basalt from the Sea of Tranquillity (Apollo 11) and a model isochron for soil samples. The basalts form two distinct clusters, as discussed in the text. The line joining these two clusters is a meaningful isochron when it can be shown that all of the basalts did, in fact, have the same initial $^{87}Sr/^{86}Sr$ ratio and the same age. Independent mineral isochrons for these rocks have been obtained and do support this assumption in this case. The two soil samples clearly do not fit the whole-rock basalt isochron, but fit a model isochron of 4.67 billion years relative to ADOR. The apparent presence of excess radiogenic ^{87}Sr in the soil compared to the basalts requires the presence of a "magic component." ● = whole-rock samples; × = soil samples. (Adapted from PAPANA-STASSIOU, WASSERBURG, and BURNETT, 1970)

The results of age determinations of basalts from the Ocean of Storms (Apollo 12) are somewhat different. In the first place, the rocks resemble the low-potassium basalts (Group II) of Tranquillity Base and have low Rb/Sr ratios. Eight basalt specimens from the Ocean of Storms were dated by PAPANASTASSIOU and WASSERBURG (1970, 1971) and by Lunatic Asylum (1971). Internal mineral isochrons indicate dates ranging from $(3.16 \pm 0.09) \times 10^9$ to $(3.36 \pm 0.10) \times 10^9$ years, as shown in Table XI.1. The basalts from the Ocean of Storms are significantly younger than those of Tranquillity Base, and thus extend the duration of volcanic activity on the Moon to the more recent past. The age determinations of lunar basalts therefore establish the important fact that volcanic activity occurred on the Moon during its early history at least until about 3 billion years ago. It is not yet clear whether the volcanic activity was due to internal heating of the Moon or due to meteorite impacts.

The basalts from the Ocean of Storms and the low-potassium basalts of Tranquillity Base have model ages of about 4.6 billion years with respect to BABI. Fig. XI.3 is a plot of the Apollo 12 basalts on an isochron diagram to show their fit to a 4.6-billion-year whole-rock isochron, in marked contrast to their individual internal mineral isochrons which indicate that crystallization occurred about 3.3 billion years ago. This phenomenon was discussed by PAPANASTASSIOU and WASSERBURG (1970, 1971). It means that

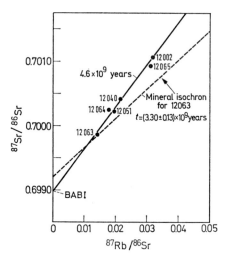

Fig. XI.3. Whole-rock isochron for samples of basalt from the Ocean of Storms (Apollo 12). It is apparent that these rocks fit a 4.6-billion-year isochron fairly well. An internal mineral isochron for sample 12063 is also shown to emphasize the fact that all of these basalts crystallized about 3.3 billion years ago. The significance of this discrepancy is mentioned in the text and has been discussed by PAPANASTASSIOU and WASSERBURG (1971), from whom these data were taken

the magmas from which these basalts crystallized were formed virtually without fractionation of rubidium and strontium during melting of rocks in the interior of the Moon and that their magma sources or reservoirs became closed systems about 4.6 billion years ago, at least insofar as rubidium and strontium are concerned. The high-potassium basalts from Tranquillity Base do not show this and seem instead to have been enriched in rubidium relative to strontium during magma formation. This point is further illustrated in Fig. XI.4.

3. Initial $^{87}Sr/^{86}Sr$ Ratios

The initial $^{87}Sr/^{86}Sr$ ratios of the basalt samples from the Moon provide information regarding the evolutionary history of lunar material. The first observation to be made from the data in Table XI.1 is that the initial ratios of the lunar basalts are all somewhat higher than BABI, which can be used as a reference point relative to which the initial $^{87}Sr/^{86}Sr$ ratios of lunar material can be evaluated. Let us agree that the initial $^{87}Sr/^{86}Sr$ ratios of the lunar basalts correctly indicate the isotopic composition of strontium of their source regions in the interior of the Moon at the time of crystallization of the basalt. If this premise is accepted, then it is possible to specify a hypothetical model for the evolution of strontium in the Moon: The Moon separated from the solar nebula at very nearly the same time as the achondrites and at that time contained strontium of uniform isotopic composition equal to BABI. Subsequently, the $^{87}Sr/^{86}Sr$ ratios in the interior of the Moon diverged because of regional differences of the Rb/Sr ratio. Basalt magma that was generated at different times, or in different places at the same time, inherited the $^{87}Sr/^{86}Sr$ ratio of its parent material. After crystallization of the basalt magma at or near the surface of the Moon, the strontium evolved at different rates depending on the Rb/Sr ratio of each specimen. This sequence of events is shown in Fig. XI.4 for basalt samples from Tranquillity Base.

The observed differences in the initial $^{87}Sr/^{86}Sr$ ratios of basalt samples from the Sea of Tranquillity indicate that these rocks were derived from at least two, and possibly several, sources having different Rb/Sr ratios. The absolute values of this ratio for the parent material can be estimated, using Eq. (II.16) and assuming single-stage histories. Taking samples 10017 and 10044 as being representative of the two major suites of basalt, one finds Rb/Sr ratios ranging from 0.003 to 0.009. The Rb/Sr ratio of the Moon is therefore significantly lower than the chondrite value (Rb/Sr ~ 0.3). Moreover, it is clear that the separation of the Moon from the solar nebula (Rb/Sr ~ 0.6) was accompanied by a drastic reduction in the Rb/Sr ratio.

The initial $^{87}Sr/^{86}Sr$ ratios of basalts from the Ocean of Storms appear to be somewhat higher than those from Transquillity Base. This is in accord with their more recent mineral isochron dates and with the suggestion that their strontium is more highly evolved because it was associated with rubidium in the interior of the Moon for a longer period of time.

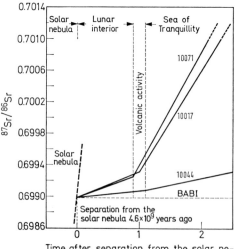

Fig. XI.4. Hypothetical model for the evolution of strontium in lunar basalt from
the Sea of Tranquillity. The model is based on the assumption that the Moon
separated from the solar nebula 4.6 × 10⁹ years ago and that lunar strontium at
that time had a ⁸⁷Sr/⁸⁶Sr ratio equal to BABI. Subsequently, the strontium in the
lunar interior evolved at different rates because of regional variation of the Rb/Sr
ratio. Approximately 1 billion years after separation from the solar nebula, basalt
magmas were generated in the interior of the Moon below the present location of
the Sea of Tranquillity. After crystallization of the basalt at or near the surface of
the Moon, the evolution of strontium continued at differing rates depending on
the Rb/Sr ratios of the basalt. Note that the separation of the Moon from the
solar nebula resulted in a sharp decrease in the Rb/Sr ratio and that the basalts
generally have higher Rb/Sr ratios than their source regions in the interior of the
Moon. (Data from PAPANASTASSIOU, WASSERBURG, and BURNETT, 1970)

4. Luny Rock 1 and Sample 12013

In the course of study of returned lunar samples two unusual rock
specimens have turned up. One of these is Luny Rock 1 (sample 10085)
from the Sea of Tranquillity. It was described by ALBEE et al. (1970, p. 464)
as a fine-grained rock fragment (shocked) consisting of low-calcium
pyroxene, isotropic "plagioclase" and "potassium-feldspar," ilmenite,
troilite, chlorofluorapatite, and whitlockite. It contains 0.56 percent potassium, 16.7 ppm rubidium, and has a ⁸⁷Sr/⁸⁶Sr ratio of 0.71399 ± 0.00011.
A date of 4.44 billion years can be calculated for this rock relative to BABI.
The authors suggested that this peculiar rock may represent non-mare
material.

The other peculiar rock is sample 12013 from the Ocean of Storms. It
was studied very carefully and the results were published in one issue of

Earth and Planetary Science Letters (Vol. 9, No. 2, September 1, 1970, 94—215, including colored photographs). The unusual nature of this rock (size: 4 × 3 × 2 cm, weight: 32.2 g) was first revealed by the results of gamma-ray spectrometry of the total sample, which showed 2.02 percent potassium, 34.3 ppm thorium, and 10.7 ppm uranium. These concentrations are 40 times higher than those of typical basalt from the Ocean of Storms. Its potassium content is 10 times that of the high-potassium basalt from Tranquillity Base (ANDERSON, 1970). DRAKE et al. (1970, p. 103) summarized the mineralogy and petrology of this rock as follows: "The rock is extremely heterogeneous and consists of at least two separate fragmental units (light and dark) permeated by a once-fluid granitic component. The fragmental material includes a wide variety of lithic and crystal fragments some of which have not yet been reported from other lunar samples. The granitic component is essentially bimineralic, with dominant potassic feldspar plus silica."

The Rb-Sr decay systematics in several fragments and density fractions of sample 12013 were investigated by Lunatic Asylum (1970). They found that lithic fragments from this rock formed about 4.6 billion years ago (whole-rock isochron), but were internally homogenized $(4.00 \pm 0.05) \times 10^9$ years ago (mineral isochrons). The whole-rock date is evidence of the oldest magmatic activity on the Moon. The initial $^{87}Sr/^{86}Sr$ ratios of the internal mineral isochrons are significantly higher than BABI and range upward to values greater than 0.708.

5. The Soil

The soil on the Moon consists of a mixture of a great variety of rock and mineral particles, most of which appear to be locally derived. However, it also contains an exotic component, including much glassy material, which may have originated elsewhere and been spread about the surface of the Moon by repeated meteorite impacts.

The model dates calculated from analyses of soil samples have invariably exceeded those obtained for the local basalts (see Fig. XI.2). For example, PAPANASTASSIOU, WASSERBURG, and BURNETT (1970) reported a date of 4.67 billion years for soil from Tranquillity Base, whereas soil from the Ocean of Storms has a date of 4.44 billion years, according to PAPANASTASSIOU and WASSERBURG (1970). Both dates were calculated relative to ADOR, the initial $^{87}Sr/^{86}Sr$ ratio of the achondrite Angra dos Reis, and are therefore somewhat higher than model dates calculated relative to BABI.

The dates derived from the soil samples present a strange anomaly because the soil appears to be older than the basaltic rocks which make up a major fraction of its bulk. Because the soil is known to be a mixture of different kinds of material, these dates are *fictitious*. However, they do suggest the existence of a "magic component" which dominates the radiogenic ^{87}Sr content of the soil, either because it is very much older than the basalt or because it has a high rubidium content relative to strontium, in addition to being older.

This "magic component" has not yet been isolated. However, there are two possibilities. One of these is the "granite" phase of sample 12013 from the Ocean of Storms. The other is a yellow-brown glass which was identified in soil samples from the Ocean of Storms by MEYER et al. (1971). This glass is characteristically enriched in potassium, the rare earth elements, and phosphorus, and is consequently referred to as KREEP. HUBBARD, GAST, and MEYER (1971) suggested that KREEP and similar material in Luny Rock 1 and sample 12013 define a distinctly different and widespread rock type on the Moon which has many of the properties inferred for the "magic component" of the soil.

6. Luna 16

On September 20, 1970, the unmanned space probe Luna 16 of the U.S.S.R. landed in the Sea of Plenty at 0°41'S. latitude and 56°18'E. longitude at a spot approximately 100 km west of Webb crater. The probe collected a core sample (35 cm long, weighing 101 g) which was returned to Earth for analysis by Soviet scientists. VINOGRADOV (1971) presented a summary of their findings during the second Lunar Conference on January 11 to 14, 1971, in Houston, Texas.

The core sample was found to consist primarily of fine dust (average grain size about 0.1 mm) of basaltic rocks and glassy material. Age determinations by the Rb-Sr method of crystalline rocks and of the fine-grained fraction of the sample were carried out. The crystalline rock was said to have an isochron date of 4.45 billion years, while the fines were dated as 4.65 ± 0.5 billion years. These preliminary results for the fines are in accord with dates for the soil from the Sea of Tranquillity and the Ocean of Storms.

7. Summary

Age determinations of basaltic rocks from the Moon disclose the occurrence of volcanic activity on the Moon more than one billion years after its formation. This indicates that the Moon experienced a period of internal geochemical differentiation early in its history before becoming extinct. The age of basalt samples from Tranquillity Base is 3.65 ± 0.06 billion years, whereas those from the Ocean of Storms have ages ranging from 3.16 to 3.36 billion years. The presence of peculiar rocks such as Luny Rock 1 and sample 12013 suggests that magmatic differentiation on the Moon produced residues approaching granitic compositions. The anomalously old dates obtained for lunar soil require the presence of a "magic component" with an unusually high content of radiogenic ^{87}Sr. The presence of recently-discovered KREEP glass or of granitic material such as that of sample 12013 may explain this anomaly.

XII. The Evolution of the Isotopic Composition of Terrestrial Strontium

1. Introduction

Many thousands of analyses of the isotopic composition of strontium in a great variety of terrestrial rocks have been made. These data, together with recent work on meteorites and lunar rocks, can provide the basis for a synthesis of the isotopic evolution of strontium within the Earth. Some aspects of this have already been presented in Chapter III and elsewhere in this book. In the following sections of this chapter we shall review the evidence and summarize the present state of these discussions. The reader may have serious reservations about the models to be described and is free to use his own judgment regarding their plausibility. It seems desirable to us, nevertheless, to end this book with speculations about the larger implications of the isotopic compositions of strontium in the rocks of the Earth.

2. Evolution of Strontium in the Mantle

Recent work on stony meteorites, summarized in Chapter X, strongly suggests that planetary objects separated from the solar nebula within a short interval of time about 4.6 billion years ago. The isotopic composition of strontium in the solar system at that time is most precisely indicated by basaltic achondrites, which had an initial $^{87}Sr/^{86}Sr$ ratio of about 0.699. This value is not necessarily representative of the strontium that was initially incorporated into all of the parent bodies of meteorites or the planets. However, there is a long-standing tradition for the analogy between the Earth and meteorites, and most scientists agree that it is valid at least as a first approximation. We therefore begin the discussion of the isotopic evolution of terrestrial strontium by specifying that the "primordial" $^{87}Sr/^{86}Sr$ ratio of the Earth 4.6 billion years ago had a value of 0.69898 (BABI), which we round up to 0.699.

Because little is known about the initial internal structure of the Earth, we assume that the mantle formed very soon after accretion of the proto-Earth. The upper part of the mantle, extending to a depth of about 600 km, subsequently produced the crust of the Earth by upward concentration of silica, alumina, and alkali metals. The evolutionary history of strontium in the upper mantle under the continents, therefore, involves not only the decay of ^{87}Rb to ^{87}Sr, but may also have been affected by time-dependent changes of the Rb/Sr ratio, resulting from transport of rubidium from the upper mantle into the crust. To complicate matters further, we recall that the variation of the $^{87}Sr/^{86}Sr$ ratios of young basaltic rocks indicates that the Rb/Sr ratio of the upper mantle is characterized by lateral and/or vertical variations. This means that we cannot adequately represent the isotopic evolution of its strontium by a single development line. Nevertheless, we do so in the following discussion in order to outline several alternative paths for the evolution of strontium in the upper mantle.

According to information presented in Chapter IV, the average present-day $^{87}Sr/^{86}Sr$ ratio of basalts in oceanic islands is about 0.7037. This value is representative of the magma sources in the upper mantle below oceanic islands. The $^{87}Sr/^{86}Sr$ ratios of young basaltic rocks on the continents appear to be slightly higher, but it is not clear whether this is due to contamination with foreign strontium from the crust or whether it indicates a real difference in the isotopic composition of strontium in the upper mantle under the continents. On the other hand, the $^{87}Sr/^{86}Sr$ ratios of oceanic tholeiites along mid-ocean ridges are significantly lower than 0.7037. These differences illustrate the isotopic heterogeneity of strontium in the upper mantle. Nevertheless, we postulate that primordial strontium 4.6 billion years ago had an $^{87}Sr/^{86}Sr$ ratio of 0.699 and that the source regions of basalt magma under oceanic islands now have an average $^{87}Sr/^{86}Sr$ ratio of 0.7037.

There are an infinite number of ways by which strontium in the upper mantle could have evolved to its present isotopic composition. Three possible paths are shown in Fig. XII.1. Others have been discussed by HART (1969) and HART and BROOKS (1970).

Case 1 in Fig. XII.1 assumes that the $^{87}Sr/^{86}Sr$ ratio of the upper mantle increased along a straight line and that its Rb/Sr ratio has remained constant.

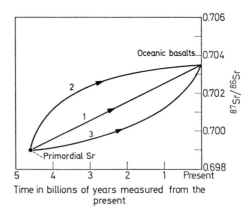

Fig. XII.1. Possible evolutionary paths for strontium in the upper mantle below oceanic islands. The histories implied by the three cases shown on this strontium development diagram are discussed in the text

This case was used in Chapter III to suggest that the upper mantle has an average, time-integrated Rb/Sr ratio of about 0.025. The requirement of a constant Rb/Sr ratio can be met in two plausible ways: (1) Rubidium and strontium were removed from the upper mantle in equal proportions, leaving the Rb/Sr ratio unchanged. (2) The upper mantle remained a closed system until magma was generated by melting and successive generations of magma were withdrawn from different regions which were sampled only once.

In Case 2 the Rb/Sr ratio of the upper mantle decreased continuously from some higher initial value to a lower present value. The decrease of the Rb/Sr ratio can be attributed to preferrential removal of rubidium from the upper mantle relative to strontium.

Case 3 requires that the Rb/Sr ratio of the upper mantle increased continuously and implies that its present value may be a maximum. There is no plausible mechanism to account for this type of behavior — which suggests that Case 3 has little validity.

We are therefore left with two possible evolutionary models represented by Case 1 and Case 2. It is not yet possible to make an unequivocal choice between these two alternatives. However, it will be instructive to examine some pertinent data which may ultimately permit us to make a choice.

3. Initial ⁸⁷Sr/⁸⁶Sr Ratios of Gabbro and Basalt

We may be able to determine whether the isotopic evolution of strontium in the upper mantle has been linear or non-linear by measuring initial ⁸⁷Sr/⁸⁶Sr ratios of mantle-derived rocks of different ages. If the rocks chosen for this purpose remained closed to rubidium and strontium after crystallization and were formed from magma derived from the upper mantle without contamination with foreign strontium, their initial ⁸⁷Sr/⁸⁶Sr ratios should define a set of development lines for strontium in the upper mantle. Several scientists have measured initial ratios of large, differentiated gabbroic intrusives with this objective in mind, and some of the resulting data were discussed in Chapter VII. FAURE et al. (1962) and FENTON and FAURE (1969) concluded on the basis of their data that the evolutionary path of strontium in the upper mantle has been strongly nonlinear, and they suggested that rubidium had been preferentially removed from the upper mantle in early Precambrian time prior to about 3 billion years ago. DAVIES et al. (1970) measured the ages and initial ⁸⁷Sr/⁸⁶Sr ratios of mafic intrusives in South Africa. They interpreted their data in terms of a two-stage model in which the Rb/Sr ratio of the upper mantle is assumed to have changed from 0.10 to 0.03 about 3.5 billion years ago.

In spite of the efforts that have been made to define the evolutionary paths of strontium in the upper mantle by the method discussed above, we are still far from a satisfactory solution. The most serious difficulty is the requirement that the magmas have not been contaminated with radiogenic ^{87}Sr during intrusion into the crust, or subsequent to crystallization. This assumption is difficult to evaluate because the contamination of strontium in a basaltic magma is not necessarily accompanied by assimilation of granitic country rocks which would cause a detectable change in the bulk chemical composition of the magma. For example, DAVIES et al. (1970) reported anomalously high initial ^{87}Sr/^{86}Sr ratios for the mafic rocks of the Bushveld Complex and the Losberg Complex in South Africa, as did PANKHURST (1969) for gabbroic rocks from Scotland (see Chapter VII). Such anomalies can be attributed either to assimilation of crustal material, selective diffusion of ^{87}Sr, isotopic equilibrium with a source enriched in radiogenic ^{87}Sr, or inhomogeneities in the upper mantle. Only when contamination with foreign strontium can definitely be ruled out can we trust these rocks to transmit information about the upper mantle. Another practical problem concerns interlaboratory discrepancies of measured ^{87}Sr/^{86}Sr ratios which result both from procedural and instrumental differences. These discrepancies are generally less than 0.0010, but they become very important in this context.

With these limitations in mind, we have compiled a list of initial ^{87}Sr/^{86}Sr ratios and ages of some mafic and alkalic igneous rocks from continental areas. Because of their mafic character, it is reasonable to assume that they were derived from the upper mantle. They also represent large volumes of magma and therefore may have been comparatively insensitive to contamination with foreign strontium from the granitic rocks of the continental crust. The data in Table XII.1 have been plotted in Fig. XII.2 in

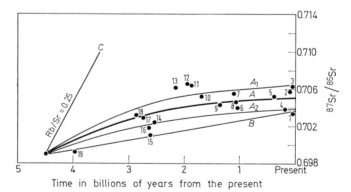

Fig. XII.2. Isotopic evolution of strontium in the upper mantle. The data points are from Table XII.1 and are identified by number. Error bars have been omitted to avoid obscuring the diagram. Curves A_1 and A_2 are envelopes which include most of the data points. Curve A is an attempted best fit midway between A_1 and A_2. Although the scatter is considerable, nonlinear evolution of strontium in the upper mantle under the continents seems to be indicated

Table XII.1. Age and initial $^{87}Sr/^{86}Sr$ ratios of mafic igneous rocks of assumed mantle origin

	Locality	Age in millions of years	$^{87}Sr/^{86}Sr$[a]	Reference
1.	Columbia River Basalt, Oregon, USA	15—25	0.7034 ± 0.0003	FENTON, unpub.
2.	Deccan Basalt, India	40—70	0.7058 ± 0.0002	FENTON, unpub.
3.	Skaergaard Intrusive, Greenland (border facies)	52	0.7063	HAMILTON (1964)
4.	Rainy Creek Complex, Montana, USA	185 ± 3	0.7038 ± 0.0002	FENTON and FAURE (1970)
5.	Garabal Hill-Glen Fyne, Igneous Complex, Scotland	392 ± 4	0.7052 ± 0.002	SUMMERHAYES (1966)
6.	Port Coldwell Complex, Ontario, Canada	1052 ± 15	0.7040 ± 0.0005	CHAUDHURI, BROOKINS, and FENTON (1971)
7.	Duluth Gabbro, Minnesota, USA	1115 ± 14	0.7055 ± 0.0003	FAURE, CHAUDHURI, and FENTON (1969)
8.	Endion Sill, Minnesota, USA	1092 ± 15	0.7046 ± 0.0006	FAURE, CHAUDHURI, and FENTON (1969)
9.	Trompsburg Intrusive, South Africa	1372 ± 142	0.7043 ± 0.0004	DAVIES et al. (1970)
10.	Norite Complex, Sudbury, Ontario	1704 ± 19	0.7052 ± 0.0003	FAIRBAIRN et al. (1967)
11.	Losberg Intrusive, South Africa	1881 ± 282	0.7064 ± 0.0024	DAVIES et al. (1970)
12.	Bushveld Complex, South Africa	1954 ± 30	0.7065 ± 0.0020	DAVIES et al. (1970)
13.	Nipissing Diabase, Ontario, Canada	2162 ± 27	0.7061 ± 0.0005	FAIRBAIRN et al. (1969)
14.	Great Dyke, Southern Rhodesia	2541 ± 30	0.7024 ± 0.0008	DAVIES et al. (1970)
15.	Modipe Gabbro, South Africa	2630 ± 470	0.7010 ± 0.0010	McELHINNY (1966)
16.	Archaean Meta-Volcanics, Canada	2600—2700	0.7018 ± 0.0002	HART and BROOKS (1970)
17.	Stillwater Complex, Montana, USA	2750[b]	0.7029 ± 0.0006	FENTON and FAURE (1969)
18.	Usushwana Complex, South Africa	2874 ± 30	0.7031 ± 0.0028	DAVIES et al. (1970)
19.	Amitsoq Gneisses, Godthaab, West Greenland	3980 ± 170	0.6992 ± 0.0005	BLACK et al. (1971)

[a] Not corrected for interlaboratory discrepancies.
[b] Date from NUNES and TILTON (1971).

order to attempt to set limits on the possible evolutionary history of strontium in the upper mantle under the continents.

It is evident that the points scatter widely and that they certainly do not define a single development line. It is probably significant, however, that most of the points lie significantly above the straight line (*B*), which represents the case of a constant Rb/Sr ratio in the upper mantle under oceanic islands. Curves A_1 and A_2 have been drawn so as to include most of the data points, but exclude the Losberg Intrusive, the Bushveld Complex, and the Nipissing Diabase — all of which have anomalously high intial $^{87}Sr/^{86}Sr$ ratios — and the Amitsoq Gneisses of Greenland — which appear to be very old and have a very low initial $^{87}Sr/^{86}Sr$ ratio. Curve A is a compromise between A_1 and A_2 and is a reasonably good fit to all of the data points with the exception of those mentioned above. Line C is the development line for chondritic material and probably represents an upper limit for the Rb/Sr ratio of the Earth. It is evident from the scatter of the points in Fig. XII.2 that the data are not sufficient to provide a unique solution. We

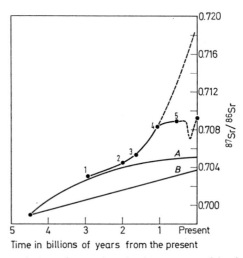

Fig. XII.3. Possible evolution of strontium in the oceans and in the continental crust. Lines *A* and *B* are from Fig. XII.2 and represent alternative models for evolution of strontium in the upper mantle under the continents. The diagram also shows measurements of $^{87}Sr/^{86}Sr$ ratios of marine carbonate rocks plotted versus their ages. The data points are: (1) Bulawayan, Limestone, Southern Rhodesia, 2950 million years, 0.7030; (2) Transvaal Limestone, South Africa, 2000 million years, 0.7044; (3) Gunflint Formation, Ontario, 1650 million years, 0.7053; (4) Nonesuch Shale (Junior), Michigan, 1075 million years, 0.7083; (5) Leverett Formation, Antarctica, 550 million years, 0.7089. The data suggest an exponential increase of the $^{87}Sr/^{86}Sr$ ratio in the oceans from about 3 to 1 billion years ago. In late Precambrian and Phanerozoic time the $^{87}Sr/^{86}Sr$ ratio of the oceans deviated sharply from the suggested exponential pattern, presumably because of increased recycling of strontium by diagenesis and weathering of marine carbonate rocks. Extrapolation of the initial exponential rise of the $^{87}Sr/^{86}Sr$ ratio in the oceans suggests that the present value of this ratio in the continental crust is about 0.718, which is consistent with other estimates discussed in Chapter III. (Data were taken from Fenton and Faure, 1969)

do believe, however, that the available evidence favors nonlinear strontium evolution in the upper mantle under the continents.

It seems to us that direct measurements of initial $^{87}Sr/^{86}Sr$ ratios of mantle-derived rocks of all ages need to be continued. Progress toward the solution of this problem will require the kind of intense effort now being devoted to the study of lunar rocks and meteorites. We hope that the lessons learned in the study of the other planets will ultimately revive interest in our own planet.

The upward migration of rubidium from the upper mantle and its concentration in the continental crust may be detectable by the consequent increase of the $^{87}Sr/^{86}Sr$ ratio in marine carbonate rocks of Precambrian age. We pointed out in Chapter VIII that the isotopic composition of strontium in the oceans is a mixture of different varieties of strontium which are released to the oceans primarily as a result of chemical weathering of rocks on the continents and in the ocean basins. The enrichment of the continental rocks in rubidium and its decay to ^{87}Sr may therefore be reflected by a rise in the $^{87}Sr/^{86}Sr$ ratio of the oceans. Marine carbonate rocks preserve this ratio virtually unchanged because of their low Rb/Sr ratios, provided they have remained closed systems. Fig. XII.3 is based on measurements of $^{87}Sr/^{86}Sr$ ratios of a few carbonate rocks and illustrates their interpretation as suggested by FENTON and FAURE (1969).

4. Continental Growth

The net loss of rubidium relative to strontium in the upper mantle under the continents that is implied by the apparent nonlinearity of the strontium development lines is indirect evidence for the growth of continents. The geochemical coherence of rubidium, potassium, silica, and alumina is well-known. One can, therefore, interpret loss of rubidium from the upper mantle as indirect evidence of the formation of sialic material and consequent growth of the continental crust.

There are two schools of thought regarding the question of continental accretion. The point of disagreement is whether the continents have grown in volume continuously throughout geologic times or whether they were formed in early Precambrian time and have not increased appreciably since then. The case for continental accretion has been supported by the discovery of "orogenic belts" on the continents, defined primarily on the basis of K-Ar age determinations of biotite. These orogenic belts appear to have been formed along the margins of continental nuclei and are believed to represent new sialic material derived directly or indirectly from the upper mantle. Fig. XII.4 attempts to show the geographic distribution of orogenic belts of North America in somewhat generalized form.

The case for continued and even accelerating growth of the continental crust has been made most presuasively by HURLEY et al. (1962, 1963), HUR-

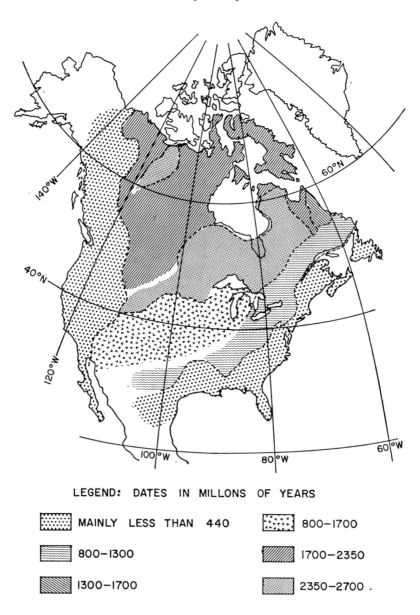

LEGEND: DATES IN MILLONS OF YEARS

MAINLY LESS THAN 440   800–1700

800–1300   1700–2350

1300–1700   2350–2700 .

Fig. XII.4. Generalized map of orogenic belts in North America. (Adapted from HURLEY and RAND, 1969, with permission of the publisher)

LEY (1967, 1968a, b), and HURLEY and RAND (1969). An important point in HURLEY's case for continental growth is his observation (HURLEY et al., 1962, Fig. 1 and 5) that the initial $^{87}Sr/^{86}Sr$ ratios of granitic rocks on the continents are generally low and do not correlate with their crystallation ages. This can be interpreted as evidence that the strontium in these rocks is juvenile and has not previously resided in the crust. It follows, therefore, that new sialic material is being added to the continental crust in the form of orogenic belts and that the continents have been growing in volume. HURLEY and RAND (1969) concluded that the rate of generation of crustal material is accelerating and amounts to 600 km^3 per million years, if both lower and upper crust are included.

HURLEY's conclusion that the continents have grown in size throughout geologic time at an accelerating rate has not gone unchallenged. On the basis of the isotopic composition of lead in oceanic sediments and beach and river sands, all of which sample rocks exposed on a continental scale, PATTERSON (1964) and PATTERSON and TATSUMOTO (1964) concluded that the formation of the continental crust occurred primarily in the interval 2.5 to 3.5 billion years ago. While allowance must be made for the difference in geochemical properties of the pairs U-Pb and Rb-Sr, PATTERSON's interpretation of the isotopic composition of lead in sialic material suggests an early enrichment of uranium into a protocrust from which the present continental crust was subsequently developed. HURLEY and PATTERSON therefore appeared to reach conflicting conclusions regarding the timing and rate of internal differentiation of the Earth.

ARMSTRONG (1968) proposed a dynamic model based on isotopic compositions of both strontium and lead, and showed that reconciliation of the models of HURLEY and PATTERSON may be possible. In ARMSTRONG's model the Earth is believed to have undergone primary differentiation into a core, mantle and sialic crust prior to about 2.5 billion years ago. Subsequently, the sialic crust and the convecting upper mantle are thought to have mixed continuously by means of sea-floor spreading, subduction of oceanic crust, melting of deep-sea sediments, and isotopic equilibration of strontium and lead with mantle material. The newly-formed sialic magmas are then either intruded into the continental crust or extruded on its surface along island arcs. Such a process would appear to be enlarging the surface areas of continents, as proposed by HURLEY, but would actually return material which had previously been derived from the continental crust. As the sediment on the sea floor is dragged into the mantle and is melted, strontium and lead are isotopically equilibrated to varying degrees with mantle material. The sialic material returning to the crust therefore may have $^{87}Sr/^{86}Sr$ ratios which make it appear to be juvenile. Thus ARMSTRONG's model can account for HURLEY's observation that granitic rocks have low $^{87}Sr/^{86}Sr$ ratios and that their ages record the time they were separated from sources in the upper mantle.

Unfortunately, the issue is clouded by semantics. HURLEY and RAND (1969, p. 1239) specifically stated that "... if material has cycled through the mantle and returned to the crust but has lost its radiogenic ^{87}Sr on the way, we refer to this as new crust, ..." This provision can be justified on the basis that the isotopic composition of strontium cannot be used to distinguish between recycled sialic crust containing mantle-equilibrated strontium and new sialic material arising from the upper mantle for the first time. The conclusion of HURLEY and RAND that the volume of the sialic crust is growing at an accelerating rate is therefore dependent on their decision to treat all of the sialic material accumulating in island arcs along continental margins as being juvenile. Another ambiguity arises from the definition of the word "crust." HURLEY and RAND (1969, p. 1237, 1238) defined the term continental crust as "... the present-day material above the Mohorovičić discontinuity," which they suggest was derived from "... mantle regions enriched in trace elements and having rubidium-strontium ratios approximately the same as the average ratio for the entire earth." HURLEY and RAND therefore do recognize the possible existence of a proto-crust, which was subsequently differentiated into the present crust. Their conclusion regarding the rate of continental accretion applies to the progressive refinement of the hypothetical protocrust into the present crust. Their "protocrust" therefore seems to be synonymous with the term "upper-mantle" as used by us and by other earth scientists.

In conclusion, we advise the interested reader to study the papers by HURLEY et al. (1962), ARMSTRONG (1968), HURLEY and RAND (1969), and HART and BROOKS (1970) to make his own judgment regarding this important controversy.

5. Summary

The evolution of the isotopic composition of strontium in the Earth was strongly affected by its differentiation into a crust and mantle. The time-dependent variation of the ^{87}Sr/^{86}Sr ratio of the upper mantle under the continents was controlled not only by the decay of ^{87}Rb to ^{87}Sr, but also by changes in its Rb/Sr ratio resulting from the upward migration of rubidium into the sialic crust. The currently-available data suggest that the evolution of the ^{87}Sr/^{86}Sr ratio of the upper mantle under the continents was non-linear due to loss of rubidium relative to strontium. The resulting decrease of the Rb/Sr ratio of the upper mantle is indirect evidence for the formation of sialic crust. Estimates of the rate of continental accretion at different times in the geologic past seem to lead to conflicting interpretations, partly because of semantic difficulties.

Appendix

Initial $^{87}Sr/^{86}Sr$ Ratios and Ages of Granitic Rocks

Locality	Initial $^{87}Sr/^{86}Sr$	Age in millions of years	Ref.
Africa			
Transvaal, Old granite	0.7106	3200	[17]
South Africa, Cape granite	0.710	553	[2]
Rhodesia, Nuanetsi syncline	0.7085	177	[34]
South-West Africa, Franzfontein	0.7080	1580	[53]
Antarctica			
Wisconsin Range, Horlick Mountains, rapakivi	0.7090	629	[33]
Wisconsin Range, Horlick Mountains, post-kinematic	0.7062	490	[33]
Asia			
India, Bundelkhand and Berach granites	0.7035	2555	[59]
Japan, Funatsu granitic rocks	0.7056	176	[61]
Australia and Tasmania			
Australia, Cape York	0.737	795	[44]
Western Tasmania, Pieman granite	0.7354	356	[23]
Western Tasmania, Heemskirk granite	0.734	354	[15]
Queensland, Sybella microgranite	0.7323	1553	[74]
South Australia, Encounter Bay	0.719	485	[63]
Western Tasmania, Meredith granite	0.715	353	[23]
Southwest Australia, Albany granite	0.7118	1100	[18]
Australia, Mount Isa, Kalkadoon granite	0.7065	1760	[35]
East Queensland, Urannah complex	0.7047	288	[52]
East Queensland, granites intruding Bulgonunna volcanics	0.7045	298	[52]
Australia, Mount Isa, Wimberu granite	0.7043	1530	[35]
East Queensland, Gayndah-Proston area	0.7041	274	[52]
East Queensland, Auburn complex	0.7040	311	[52]
East Queensland, Maryborough basin	0.7035	220	[52]
Canada			
British Columbia, White Creek batholith, core	0.725	111	[25]
Nova Scotia granites	0.708	300	[11]

Table (continued)

Locality	Initial $^{87}Sr/^{86}Sr$	Age in millions of years	Ref.
British Columbia, White Creek batholith, boundary rocks	0.7077	111	[25]
[a] British Columbia granites	0.7071	56	[16]
Northwest Territory, Ross Lake granodiorite	0.707	2510	[80]
Newfoundland, Holyrood granite	0.7040	574	[40]
Ontario, Westport granite	0.704	1016	[5]
Ontario, Rainy Lake	0.7036	2520	[46]
Northwest Territory, Western granodiorite	0.7035	2610	[8]
Southeastern Manitoba, Northern granite	0.7031	2550	[24]
Manitoba, Annabel Lake pluton	0.7023	1805	[67]
Manitoba, Black Lake quartz monzonite	0.7019	2735	[62]
Ontario, Manitouwadge	0.7015	2760	[36]
Ontario, Rainy Lake	0.7012	2400	[3]
Quebec, Dauversiere stock	0.7011	2610	[72]
Northwest Territory, Southeast granodiorite	0.7011	2640	[80]
Saskatchewan, Hanson Lake area	0.7008	2446	[60]
Ontario, North Bay granite	0.700	2700	[5]
Ontario, Round Lake	0.7009	2390	[7]

Europe

Locality	Initial $^{87}Sr/^{86}Sr$	Age in millions of years	Ref.
Scotland, Glen Clova granites	0.729	389	[1]
Scotland, Ben Vuroch, Portsoy, and Windyhills granites	0.719	480	[1]
France, Mendic granite	0.7182	453	[86]
Scotland, Kemnay, Peterhead, and Bennachie granites	0.717	393	[1]
Germany, Barhalde granite	0.7162	284	[32]
Scotland, Strichen and Longmanhill granites	0.715	501	[1]
Germany, Schluchsee granite	0.7142	315	[32]
Germany, Brocken and Oker granites	0.714	282	[71]
Scotland, Skye granites[a]	0.7124	54	[14]
Germany, granite porphyry	0.7121	274	[32]
Switzerland, Monte Rosa granite	0.712	310	[69]
Scotland, Kennetenmont and Auchedly granite	0.7106	459	[47]
France, Pyrenees	0.710	300	[55]
France, Cornil granite	0.7095	317	[77]
Portugal, Older Hercynian granite	0.709	298	[70]
Ireland, Donegal granite	0.708	470	[58]
North Atlantic, Rockall Bank[a]	0.7065	60	[45]
France, Chateau-Gaillard granite	0.7061	506	[65]
Poland, Eastern Tatra Mountains	0.706	300	[26]
Ireland, Oughterard granite	0.7059	575	[10]

Table (continued)

Locality	Initial $^{87}Sr/^{86}Sr$	Age in millions of years	Ref.
Ireland, Island granite	0.7057	388	[10]
Ireland, Galway granite	0.705	384	[10]
Portugal, Younger Hercynian granite	0.7046	280	[70]
Ireland, Inish granite	0.704	404	[10]
Norway, Oslo granite	0.703	288	[50]
South America			
Argentina, Olavarria area	0.7063	1880	[66]
Venezuela, Encrucijada granite	0.7024	2111	[29]
U.S.S.R.			
Ukraine, Korosten granites	0.709	1720	[30]
Yenesiy Range, Nizhnekansk granite	0.709	535	[31]
Karelia, Karmasel'g granites	0.705	2510	[31]
Karelia, Tervus granite	0.7043	1815	[68]
Karelia, Suoyarv-region alaskite	0.704	2190	[31]
Karelia, Karelid granites	0.704	2600	[19]
Yenesiy Range, Taraksk granite	0.703	2060	[31]
Karelia, Sunsk granites	0.702	2610	[31]
Kazakhstan, Bektau-Ata alaskite	0.701	290	[31]
Yenisey Range, Kan River-region granites	0.700	2550	[31]
United States			
Massachusetts, Quincy granite	0.728	325	[41, 42]
Colorado, biotite-muscovite granite	0.721	1390	[9]
Massachusetts, Hoppin Hill granite	0.7128	514	[28]
Connecticut, Nonewaug granite	0.712	382	[79]
Massachusetts, Clinton Quad rangle	0.7115	250	[3]
Massachusetts, Chelmsford granite	0.7113	405	[43]
California, Coast Range batholith	0.7082	~ 100	[75]
Arizona, Gila County	0.7080	1330	[3]
Sierra Nevada granites	0.7073	90	[12]
Colorado, Pikes Peak granite	0.7072	104	[78]
Massachusetts, Cape Ann granite	0.707	415	[27]
California, Inyo Mountains batholith	0.7070	~ 170	[75]
Montana, Boulder batholith granodiorites (average)	0.7066	100	[51]
Colorado, Tenmile granite	0.7065	1720	[37]
Wyoming, Sherman-type granite	0.7065	1335	[6]
New Hampshire, Lebanon dome	0.706	440	[49]
New Hampshire, phase of Conway granite	0.706	110	[56]
Texas, Llano	0.706	1000	[4]
Colorado, Eldora (monzonite)[a]	0.7059	55	[3]
Massachusetts, Andover granite	0.7058	450	[43]
Kansas, Rose Dome granite	0.705	1180	[39]

Table (continued)

Locality	Initial $^{87}Sr/^{86}Sr$	Age in millions of years	Ref.
New Hampshire, phase of Conway granite	0.705	187	[56]
New Hampshire, Mascoma dome	0.705	440	[49]
Maine, Stonington granite	0.7050	341	[57]
Wyoming, Granite Mountains	0.7049	2610	[64]
Washington, Snoqualmie granodiorite[a]	0.7045	15	[3]
Maine, Oak Point granite	0.7045	357	[57]
Colorado, Southwest, younger granites	0.7043	1466	[37]
Colorado, Whitehead granite	0.7040	1675	[37]
Colorado, St. Kevin granite	0.704	1390	[20]
Missouri, South-East	0.704	1319	[54]
South Carolina, Liberty Hill	0.7039	306	[48]
Colorado, Vernal Mesa monzonite	0.7038	1480	[22]
Colorado, Sherman granite	0.7036	1410	[9]
Wyoming, Baggot Rocks granite	0.7036	2340	[6]
Colorado, Boulder Creek granite	0.7035	1700	[9]
Minnesota, Sacred Heart granite	0.7033	2400	[3]
North Carolina, Salisbury	0.7032	411	[48]
Colorado, Log Cabin granite	0.7031	1420	[9]
Colorado, central	0.703	1640	[13]
Wyoming, foliated granite	0.7026	1470	[6]
Colorado, Longs Peak — St Vrain granites	0.7025	1450	[9]
Colorado, Vernal Mesa-type granodiorite	0.702	1694	[38]
Missouri, southeast	0.702	1415	[54]
Minnesota, Sacred Heart granite	0.702	2700	[73]
Minnesota, Ortonville granite	0.7019	2400	[3]
Colorado, Pitts Meadow granodiorite	0.7017	1730	[22]
Colorado, Bakers Bridge granite	0.7012	1711	[37]
Colorado, Silver Plume-type granite	0.701	1472	[38]
Wyoming, foliated granite	0.7002	1715	[6]
Massachusetts, Ayer granite	0.700	345	[21]
Colorado, Curecanti monzonite	0.700	1420	[22]
Missouri, southeast	0.700	1318	[54]

[a] The initial $^{87}Sr/^{86}Sr$ ratio is calculated from whole-rock Rb/Sr ratio and known age.

References for Appendix

1. BELL, K.: Age relations and provenance of the Dalradian series of Scotland. Bull. Geol. Soc. Am. 79, 1167—1194 (1968).
2. ALLSOPP, H. L., KOLBE, P.: Isotopic age determinations on the Cape Granite and intruded Malmesbury sediments, Cape Peninsula, South Africa. Geochim. Cosmochim. Acta 29, 1115—1130 (1965).
3. HEDGE, C. E., WALTHALL, F. G.: Radiogenic strontium-87 as an index of geologic processes. Science 140, 1214—1217 (1963).
4. ZARTMAN, R. E.: Rubidium-strontium age of some metamorphic rocks from the Llano uplift, Texas. J. Petrol. 6, 28—36 (1965).

5. KROGH, T. E., HURLEY, P. M.: Strontium isotope variation and whole-rock isochron studies, Grenville Province of Ontario. J. Geophys. Res. **73**, 7107—7125 (1968).

6. HILLS, F. A. et al.: Precambrian geochronology of the Medicine Bow Mountains, Southeastern Wyoming. Bull. Geol. Soc. Am. **79**, 1757—1784 (1968).

7. PURDY, J. W., YORK, D.: Rb-Sr whole-rock and K-Ar mineral ages of rocks from the Superior Province near Kirkland Lake, northeastern Ontario, Canada. Can. J. Earth Sci. **5**, 699—705 (1968).

8. GREEN, D. C., BAADSGAARD, H., CUMMING, G. L.: Geochronology of the Yellowknife area, Northwest Territories, Canada. Can. J. Earth Sci. **5**, 725—735 (1968).

9. PETERMAN, Z. E., HEDGE, C.: Chronology of Precambrian events in the Front Range, Colorado. Can. J. Earth Sci. **5**, 749—756 (1968).

10. LEGGO, P. J., COMPSTON, W., LEAKE, B. E.: The geochronology of the Connemara granites and its bearing on the antiquity of the Dalradian Series. Quart. J. Geol. Soc. London **122**, 91—118 (1966).

11. FAIRBAIRN, H. W., HURLEY, P. M., PINSON, W. H.: Preliminary age study and initial Sr^{87}/Sr^{86} of Nova Scotia granitic rocks by the Rb-Sr whole-rock method. Bull. Geol. Soc. Am. **75**, 253—258 (1964).

12. HURLEY, P. M., BATEMAN, P. C., FAIRBAIRN, H. W., PINSON, W. H., JR.: Investigation of initial Sr^{87}/Sr^{86} ratios in the Sierra Nevada plutonic province. Bull. Geol. Soc. Am. **76**, 165—174 (1965).

13. WETHERILL, G. W., BICKFORD, M. E.: Primary and metamorphic Rb-Sr chronology in Central Colorado. J. Geophys. Res. **70**, 4669—4686 (1965).

14. MOORBATH, S., BELL, J. D.: Strontium isotope abundance studies and rubidium-strontium age determinations on Tertiary igneous rocks from the Isle of Skye, Northwest Scotland. J. Petrol. **6**, 37—66 (1965).

15. BROOKS, C., COMPSTON, W.: The age and initial Sr^{87}/Sr^{86} of the Heemskirk granite, Western Tasmania. J. Geophys. Res. **70**, 6249—6262 (1965).

16. FAIRBAIRN, H. W., HURLEY, P. M., PINSON, W. H.: Initial Sr^{87}/Sr^{86} and possible sources of granitic rocks in southern British Columbia. J. Geophys. Res. **69**, 4893 (1964).

17. ALLSOPP, H. L.: Rb-Sr age measurements on total rock and separated-mineral fractions from the Old Granite of the Central Transvaal. J. Geophys. Res. **66**, 1499—1508 (1961).

18. TUREK, A., STEPHENSON, N. C. N.: The radiometric age of the Albany granite and the Sterling Range beds, southwest Australia. J. Geol. Soc. Australia **13**, 449—456 (1966).

19. GOROKHOV, I. M., LOBACH-ZHUCHENKO, S. B.: Determination of the age of the Karelid granites of southwestern Karelia by the Rb-Sr isochron method. Geochem. Intern. **5**, 944 (1964).

20. PEARSON, R. C., HEDGE, C. E., THOMAS, H. H., STERN, T. W.: Geochronology of the St. Kevin granite and neighboring Precambrian rocks, Northern Sawatch Range, Colorado. Bull. Geol. Soc. Am. **77**, 1109—1120 (1966).

21. ZARTMAN, R. E., SNYDER, G., STERN, T. W., MARVIN, R. F., BUCKHAM, R. C.: Implications of radiometric ages in eastern Connecticut and Massachusetts. U.S. Geol. Surv. Prof. Pap. 525 D, D1—D10 (1965).

22. HANSEN, W. R., PETERMAN, Z. E.: Basement rock geochronology of the Black Canyon of the Gunnison, Colorado. U.S. Geol. Surv. Prof. Pap. 600 C, C80—C90 (1968).

23. BROOKS, C.: The rubidium-strontium ages of some Tasmanian igneous rocks. Geol. Soc. Aust. J. **13**, 457—469 (1966).

24. TUREK, A., PETERMAN, Z. E.: Preliminary Rb-Sr geochronology of the Rice Lake-Beresford Lake area, southeastern Manitoba. Can. J. Earth Sci. **5**, 1373—1380 (1968).

25. WANLESS, R. K., LOVERIDGE, W. D., MURSKY, G.: A geochronological study of the White Creek batholith, southeastern British Columbia. Can. J. Earth Sci. **5**, 375—386 (1968).

26. BURCHART, J.: Rubidium-strontium isochron ages of the crystalline core of the Tatra Mountains, Poland. Am. J. Sci. **266**, 895—907 (1968).

27. BOTTINO, M. L., FULLAGAR, P. D.: The effects of weathering on whole-rock Rb-Sr ages of granitic rocks. Am. J. Sci. **266**, 661—670 (1968).

28. FAIRBAIRN, H. W., MOORBATH, S., RAMO, A. O., PINSON, W. H., JR., HURLEY, P. M.: Rb-Sr ages of granitic rocks of southeastern Massachusetts and the age of the lower Cambrian at Hoppin Hill. Earth Plan. Sci. Lett. **2**, 321—328 (1967).

29. POSADAS, V. G., KALLIOKOSKI, J.: Rb-Sr ages of the Encrucijada granite intrusive in the Imataca Complex, Venezuela. Earth Plan. Sci. Lett. **2**, 210—214 (1967).

30. GOROKHOV, I. M.: Whole-rock Rb-Sr ages of the Korosten granites, Dnieper migmatites, and metamorphosed mafic rocks of the Ukraine. Geochem. Intern. **4**, 738—746 (1964).

31. GOROKHOV, I. M., ARTEMOV, Yu. M.: Petrological significance of primary abundance of [87]Sr in igneous and metamorphic rocks. Geochem. Intern. **3**, 8—12 (1966).

32. BROOKS, C., WENDT, I., HARRE, W.: A two-error regression treatment and its application to Rb-Sr and initial [87]Sr/[86]Sr ratios of younger Variscan granitic rocks from the Schwarzwald massif; southwest Germany. J. Geophys. Res. **73**, 6071—6084 (1968).

33. FAURE, G., HILL, R. L., EASTIN, R., MONTIGNY, R. J. E.: Age determination of rocks and minerals from the Transantarctic Mountains. Antarctic J. U.S. Sept.—Oct., 173—175 (1968).

34. MANTON, W.: The origin of associated basic and acid rocks in the Lemombo-Nuanetsi igneous province, Southern Africa, as implied by strontium isotopes. J. Petrol. **9**, 23—29 (1968).

35. RICHARDS, J. R.: Some Rb-Sr measurements on granites near Mt. Isa. Australasian Inst. Mining Met. Proc. **218**, 19—23 (1966).

36. TILTON, G. R., STEIGER, R. H.: Mineral ages and isotopic composition of primary lead at Manitouwadge, Ontario. J. Geophys. Res. **74**, 2118—2132 (1969).

37. BICKFORD, M. E., WETHERILL, G. W., BARKER, F., CHIN-Nan LEE-HU: Precambrian Rb-Sr chronology in the Needle Mountains, southwest Colorado. J. Geophys. Res. **74**, 1660—1676 (1969).

38. MOSE, D. G., BICKFORD, M. E.: Precambrian geochronology in the Unaweep Canyon, West-Central Colorado. J. Geophys. Res. **74**, 1677—1687 (1969).

39. BICKFORD, M. E., MOSE, D. G.: Age of the Rose Dome granite, Woodson County, Kansas. Geol. Soc. Am. Abstracts with Programs for 1969, Pt. 2, p. 2 (1969).

40. McCARTNEY, W. D., POOLE, W. H., WANLESS, R. K., WILLIAMS, H., LOVERIDGE, W. D.: Rb/Sr age and geological setting of the Holyrood granite, southeast Newfoundland. Can. J. Earth Sci. **3**, 947—957 (1966).

41. LYONS, J.B., FAUL, H.: Isotope geochronology of the northern Appalachians. In: Studies of Appalachian Geology: Northern and Maritime. New York: Interscience 305—318 (1968).

42. BOTTINO, M.L., PINSON, W.H., FAIRBAIRN, H.W., HURLEY, P.M.: Whole-rock Rb-Sr ages of some Paleozoic volcanics and related granites in the Northern Appalachians. Trans. Am. Geophys. Union 44, 111 (1963).

43. FAIRBAIRN, H.W., BOTTINO, M.L., HANFORD, L.S., HURLEY, P.M., HEATH, M.M., PINSON, W.H.: Radiometric ages of igneous rocks in northeastern Massachusetts. Geol. Soc. Am., Northeastern Section, Program for 1967 Annual Meeting, Boston (1967).

44. KAPLAN, G.: Analyse geochronologique de quelques roches plutoniennes de la Peninsula du Cap York (Queensland-Australie). Bull. Centre Rech. Pau SNPA 2, 399—409 (1968).

45. MOORBATH, S., WELKE, H.: Isotopic evidence for the continental affinity of the Rockall Bank, North Atlantic. Earth Plan. Sci. Lett. 5, 211—216 (1969).

46. HART, S.R., DAVIS, G.L.: Zircon U-Pb and whole-rock Rb-Sr ages and early crustal development near Rainy Lake, Ontario. Bull. Geol. Soc. Am. 80, 595—616 (1969).

47. PANKHURST, R.J.: Strontium isotope studies related to petrogenesis in the Caledonian basic igneous province of northeast Scotland. J. Petrol. 10, 115—143 (1969).

48. FULLAGAR, P.D., LEMMON, R.E., RAGLAND, P.C.: Petrochemical and geochronological studies of plutonic rocks in the Southern Appalachians: Part 1, The Salisbury Pluton. Bull. Geol. Soc. Am. 82, 409—416 (1971).

49. NAYLOR, R.S.: Age and origin of the Oliverian domes, central-western New Hampshire. Bull. Geol. Soc. Am. 80, 405—428 (1969).

50. HEIER, K.S., COMPSTON, W.: Rb-Sr isotopic studies of the plutonic rocks of the Oslo region. Lithos 2, 133—145 (1969).

51. DOE, B.R., TILLING, R.I., HEDGE, C.E., KLEPPER, M.R.: Lead and strontium isotopic studies of the Boulder batholith, southwestern Montana. Econ. Geol. 63, 884—906 (1968).

52. WEBB, A.W., McDOUGALL, I.: The geochronology of the igneous rocks of eastern Queensland. J. Geol. Soc. Austalia 15, 313—346 (1968).

53. CLIFFORD, T.N., ROOKE, J.M., ALLSOPP, H.L.: Petrochemistry and age of the Franzfontein granitic rocks of northern Southwest Africa. Geochim. Cosmochim. Acta 33, 973—986 (1969).

54. ANDERSON, J.E., BICKFORD, M.E., ODOM, A.L., BERRY, A.W.: Some age relations and structural features of the Precambrian volcanic terrane, St. Francois Mountains, southeastern Missouri. Bull. Geol. Soc. Am. 80, 1815—1818 (1969).

55. VITRAC, A., ALLEGRE, C.J.: Age de mise en place, origine et histoire des granites de Millas Querigut-Mont Louis etudies par la methode $^{87}Rb/^{87}Sr$. C.R. Acad. Sci. Paris 269, 2174—2177 (1969).

56. FOLAND, K.A., QUINN, A.W., GILETTI, B.J.: K-Ar and Rb-Sr Jurassic and Cretaceous ages for intrusives of the White Mountain magma series, northwestern New England. Am. J. Sci. 270, 321—330 (1971).

57. BROOKINS, D.G., SPOONER, C.M.: The isotopic ages of the Oak Point and Stonington granites, eastern Penobscot Bay, Maine. J. Geol. 78, 570—576 (1970).

58. Leggo, P. J., Tanner, P. W. G., Leake, B. E.: Isochron study of Donegal granite and certain Dalradian rocks of Britain, in "North Atlantic: Geology and Continental Drift." Am. Assoc. Petrol. Geol. Mem. **12**, 354—362 (1969).

59. Crawford, A. R.,: The Precambrian geochronology of Rajusthan and Bundelkhand, northern India. Can. J. Earth. Sci. **7**, 91—110 (1970).

60. Coleman, L. C.: Rb/Sr isochrons for some Precambrian rocks in the Hanson Lake area, Saskatchewan. Can. J. Earth Sci. **7**, 338—345 (1970).

61. Shibata, K., Nozawa, T., Wanless, R. K.: Rb-Sr geochronology of the Hida metamorphic belt, Japan. Can. J. Earth Sci. **7**, 1383—1401 (1970).

62. Turek, A., Peterman, Z. E.: Advances in the geochronology of the Rice Lake — Beresford Lake area, southeastern Manitoba. Can. J. Earth Sci. **8**, 572—579 (1971).

63. Dasch, E. J., Milnes, A. R., Nesbitt, R. W.: Rubidium-strontium geochronology of the Encounter Bay granites and adjacent metasedimentary rocks, South Australia. Geol. Soc. Am. Abstracts with Programs **3**, 107 (1971).

64. Peterman, Z. E., Hildreth, R. A., Nkomo, I.: Precambrian geology and geochronology of the Granite Mountains, central Wyoming. Geol. Soc. Am. Abstracts with Programs **3**, 403 (1971).

65. Cantagrel, J. M., Valizadeh, M., Vialette, Y.: Age of granites, granophyres, and kersanites of the Thiers region (Puyde-Dome) in the French Massif Central. C. R. Acad. Sci., Ser. D. **270**, 600—603 (1970).

66. Halpern, M., Linares, E.: Rubidium-strontium age of granitic rocks from the crystalline basement of the Olavarria area, Buenos Aires Province, Argentina. Rev. Asoc. Geol. Arg. **25**, 303—306 (1970).

67. Mukherjee, A. C., Stauffer, M. R., Baadsgaard, H.: The Hudsonian Orogeny near Flin Flon, Manitoba: a tentative interpretation of Rb/Sr and K/Ar ages. Can. J. Earth Sci. **8**, 939—946 (1971).

68. Gorokhov, I. M., Varshavskaya, E. S., Kutyavin, E. P., Lobach-Zhuchenko, S. B.: Preliminary Rb-Sr geochronology of the North Ladoga region, Soviet Karelia. Eclogae Geol. Helv. **63**, 95—104 (1970).

69. Hunziker, J. C.: Polymetamorphism in the Monte Rosa, Western Alps. Eclogae Geol. Helv. **63**, 151—161 (1970).

70. Priem, H. N. A., Boelrijk, N. A. I. M., Verschure, R. H., Hebeda, E. H., Verdurmen, E. A. Th.: Dating events of acid plutonism through the Paleozoic of the Western Iberian Peninsula. Eclogae Geol. Helv. **63**, 255—274 (1970).

71. Schoell, M.: K/Ar and Rb/Sr age determinations on minerals and total rocks of the Harz Mountains/Germany. Eclogae Geol. Helv. **63**, 299 (1970).

72. Wanless, R. K., Stevens, R. D., Loveridge, W. D.: Anomalous parent-daughter isotopic relationships in rocks adjacent to the Grenville front near Chibougamau, Quebec. Eclogae Geol. Helv. **63**, 345—364 (1970).

73. Goldich, S. S., Hedge, C. E., Stern, T. W.: Age of the Morton and Montevideo gneisses and related rocks, southwestern Minnesota. Bull. Geol. Soc. Am. **81**, 3671—3696 (1970).

74. Farquharson, R. B., Richards, J. R.: Whole-rock U-Th-Pb and Rb-Sr ages of the Sybella microgranite and pegmatite, Mount Isa, Queensland. J. Geol. Soc. Austalia **17**, 53—58 (1970).

75. Kistler, R. W., Evernden, J. F., Shaw, H. R.: Sierra Nevada plutonic cycle: Part I, origin of composite granitic batholiths. Bull. Geol. Soc. Am. **82**, 853—868 (1971).

76. ROQUES, M., VACHETTE, M.: Ages au strontium sur roches totales des migmatites de la zone axiale de la Montagne Noire et du massif de granite du Mendic (Massif Central Francais). C.R. Acad. Sci. Paris **270**, 275—278 (1970).

77. BERNARD-GRIFFITHS, J., VACHETTE, M.: Age cambrien de migmatites de l'Antinclinal de Tulle (Massif Central Francais) et ses relations avec l'age du granite dit "tardimigmatitique" de type Cornil. C. R. Acad. Sci. Paris **270**, 916—919 (1970).

78. HEDGE, C. E., JR: Whole-rock Rb-Sr age of the Pikes Peak Batholith, Colorado. U. S. Geol. Surv. Prof. Pap. 700-B, 86—89 (1970).

79. BESANCON, J. R.: A Rb-Sr isochron for the Nonewaug granite. Geol. Natural. Hist. Surv. Conn. Rept. Invest. 5, 1—9 (1970).

80. GREEN, D. C., BAADSGAARD, H.: Temporal evolution and petrogenesis of an Archean crustal segment at Yellowknife, N.W.T., Canada. J. Petrology **12**, 177—217 (1971).

Bibliography

AHRENS, L.: Measuring geologic time by the strontium method. Geol. Soc. Am. Bull. **60**, 217—266 (1949).

AHRENS, L. H., PINSON, W. H., KEARNS, M. W.: Association of rubidium and potassium and their abundance in common igneous rocks and meteorites. Geochim. Cosmochim. Acta **2**, 229—242 (1952).

AHRENS, L. H., TAYLOR, S. R.: Spectrochemical analysis (2 d ed.). Reading, Mass.: Addison-Wesley 1960.

ALBEE, A. L., BURNETT, D. S., CHODOS, A. A., EUGSTER, O. J., HUNEKE, J. G., PAPANASTASSIOU, D. A., PODOSEK, F. A., RUSS, G. P. II, SANZ, H. G., TERA, F., WASSERBURG, G. J.: Ages, irradiation history, and chemical composition of lunar rocks from the Sea of Tranquillity. Science **167**, 463—466 (1970).

ALDRICH, L. T., DOAK, J. B., DAVIS, G. L.: The use of ion exchange columns in mineral analysis for age determination. Am. J. Sci. **251**, 377—387 (1953a).

ALDRICH, L. T., HERZOG, L. F., DOAK, J. B., DAVIS, G. L.: Variations in strontium abundances in minerals, Part I: Mass spectrometric analysis of mineral sources of strontium. Trans. Am. Geophys. Union **34**, 457—460 (1953b).

ALDRICH, L. T., WETHERILL, G. W., TILTON, G. R., DAVIS, G. L.: The half life of ^{87}Rb. Phys. Rev. **104**, 1045—1047.

ALDRICH, L. T., HART, S. R., TILTON, G. R., DAVIS, G. L., RAMA, S. N. I., STEIGER, R., RICHARDS, J. R., GERKENS, J. S.: Ultramafic rocks of St. Paul's Islands. Carnegie Inst. Wash. Year Book **63**, 330—331 (1964).

ALDRICH, L. T., DAVIS, G. L., JAMES, H. L.: Ages of minerals from metamorphic and igneous rocks near Iron Mountain, Michigan. J. Petrol. **6**, 445—472 (1965).

ALLEGRE, C., DARS, R.: Chronologie on rubidium-strontium et granitologie. Geol. Rundschau **55**, 226—237 (1965).

ALLSOPP, H. L.: Rb-Sr measurements on total rock and separated mineral fractions from the Old Granite of the Central Transvaal. J. Geophys. Res. **66**, 1499—1508 (1961).

ALLSOPP, H. L., ROBERTS, H. R., SCHREINER, G. D. L., HUNTER, D. R.: Rb-Sr measurements on various Swaziland granites. J. Geophys. Res. **67**, 5307—5313 (1962).

ALLSOPP, H. L., KOLBE, P.: Isotopic age determinations on the Cape granite and intruded Malmesbury sediments, Cape Peninsula, South Africa. Geochim. Cosmochim. Acta **29**, 1115—1130 (1965).

ALLSOPP, H. L.: Rb-Sr and K-Ar age measurements on the Great Dyke of Southern Rhodesia. J. Geophys. Res. **70**, 977—984 (1965).

ALLSOPP, H. L., BURGER, A. J., VAN ZYL, C.: A minimum age for the Premier Kimberlite pipe yielded by biotite Rb-Sr measurements, with related galena isotopic data. Earth Planet. Sci. Lett. **3**, 161—166 (1967).

ALLSOPP, H. L., ULRYCH, T. J., NICOLAYSEN, L. O.: Dating some significant events in the history of the Swaziland System by the Rb-Sr isochron method. Can. J. Earth Sci. **5**, 605—619 (1968).

ALLSOPP, H. L., NICOLAYSEN, L. O., HAHN-WEINHEIMER, P.: Rb/K ratios and Sr-isotopic compositions of minerals in eclogitic and peridotitic rocks. Earth Planet. Sci. Lett. **5**, 231—244 (1969).

AL-RAWI, Y., CARMICHAEL, I. S. E.: Natural fusion of granites. Am. Mineralogist 52, 1806—1814 (1967).

AMIRKHANOFF, K. I., BRANDT, S. B., BARTNITSKII, E. N.: Diffusion of radiogenic argon in feldspars. Dokl. Akad. Sci. U.S.S.R. 125, No. 6, 125 (1959).

AMIRKHANOFF, K. I., BRANDT, S. B., BARTNITSKII, E. N.: Radiogenic argon in minerals and its migration. Ann. N. Y. Acad. Sci. 91, 235—275 (1961).

ANDERS, E., GOLES, G. G.: Theories on the origin of meteorites. J. Chem. Educ. 38, 58—66 (1961).

ANDERS, E.: Meteorite ages. Rev. Mod. Phys. 34, 287—325 (1962).

ANDERS, E.: Meteorite ages. In: The Solar System, vol. 4, The Moon, Meteorites, and Comets, ed. by MIDDLEHURST, B. M., KUIPER, G. P.: Chicago: University of Chicago Press 1963.

ANDERS, E.: Origin, age and composition of meteorites. Space Sci. Rev. 3, 583—714 (1964).

ANDERSON, D. H.: The preliminary examination and preparation of lunar sample 12013. Earth Planet. Sci. Lett. 9, 94—102 (1970).

ARMSTRONG, R. L.: A model for the evolution of strontium and lead isotopes in a dynamic earth. Rev. Geophys. 6, 175—199 (1968).

ARMSTRONG, R. L., COOPER, J. A.: Lead isotopes in island arcs. Abstracts, Symposium on volcanoes and their roots, Oxford, IAVCEI, p. 117 (1969).

ARMSTRONG, R. L., DASCH, E. J., NESBITT, R., SCOTT, R. B.: Significance of strontium isotope evolution to the origin of Great Basin ignimbrites, U.S.A. Abstracts, Symposium on volcanoes and their roots, Oxford, IAVCEI, pp. 118—119 (1969).

ARMSTRONG, R. L.: Glacial erosion and the variable isotopic composition of strontium in seawater. Nature 230, No. 14, 132—133 (1971).

ARRIENS, P. A., BROOKS, C., BOFINGER, V. M., COMPSTON, W.: The discordance of mineral ages in granitic rocks resulting from the redistribution of rubidium and strontium. J. Geophys. Res. 71, 4981—4994 (1966).

ARTEMOV, YU. M., YAROSHEVSKIY, A. A.: The isotopic composition of strontium as an indicator of character and duration of magmatic differentiation. Geochem. Intern. 2, 810—813 (1965).

BAADSGAARD, H., VAN BREEMEN, O.: Thermally-induced migration of Rb and Sr in adamellite. Eclogae geol. Helv. 63, 31—44 (1970).

BACKLUND, H. G.: On the mode of intrusion of deep-seated alkaline bodies. Bull. Geol. Inst. Univ. Uppsala 24, 1—24 (1932).

BAILEY, D. K.: Isotopic composition of strontium in carbonatites. Nature 201, 599 (1964).

BARBERI, F., BORSI, S., FERRARA, G., INNOCENTI, F.: Strontium isotope composition of some recent basic volcanites of the S. Tyrrhenian Sea and Sicily Channel. Contr. Min. Petrol. 23, 157—172 (1969).

BARBERI, F., BORSI, S., FERRARA, G., MARINELLI, G., VARET, J.: Relationships between tectonics and magmatology in the Northern Danakil depression, Ethiopia. Phil. Trans. Roy. Soc. London (1970). In press.

BARGHOORN, E. S., SCHOPF, J. W.: Microorganisms three billion years old from the Precambrian of South Africa. Science 152, 758—763 (1966).

BARKER, D. S., LONG, L. E.: Feldspathoidal syenite in a quartz diabase sill, Brookville, New Jersey. J. Petrol. 10, 202—221 (1969).

BARKER, D. S.: North American feldspathoidal rocks in space and time. Geol. Soc. Am. Bull. 80, 2369—2372 (1969).

BARKER, D. S.: North American occurrences of alkalic rocks. In: SORENSEN, H. (ed.): The alkaline rocks. Wiley-Interscience (1972) (in press).

BARRETT, P. J., FAURE, G.: Strontium isotope composition of non-marine carbonate rocks from the Beacon supergroup of the Transantarctic Mountains, Geochim. Cosmochim. Acta (1972). (In press).

BELL, K.: Age relations and provenance of the Dalradian series of Scotland. Geol. Soc. Am. Bull. **79**, 1167—1194 (1968).

BELL, K., POWELL, J. L.: Strontium isotopic studies of alkalic rocks: The potassium-rich lavas of the Birunga and Toro-Ankole regions, East and Central Equatorial Africa. J. Petrol. **10**, 536—572 (1969).

BELL, K., POWELL, J. L.: Strontium isotopic studies of alkalic rocks: The alkalic complexes of Eastern Uganda. Geol. Soc. Am. Bull. **81**, 3481—3490 (1970).

BENCE, A. E.: The differentiation history of the Earth by rubidium-strontium isotopic relationships. In: M.I.T. Ann. Prog. Rept. 35—78 (1966).

BICKFORD, M. E., WETHERILL, G. W., BARKER, F., LEE-HU, CHIN-NAN: Precambrian Rb-Sr chronology in the Needle Mountains, Southwestern Colorado. J. Geophys. Res. **74**, 1660—1676 (1969).

BIKERMAN, M.: Isotopic studies in the Roskruge Mountains, Pima County, Arizona. Geol. Soc. Am. Bull. **78**, 1029—1036 (1967).

BISCAYE, P. E., DASCH, E. J.: The rubidium, strontium, strontium isotope system in deep sea sediments: Argentine basin. J. Geophys. Res. **76**, 5087—5096 (1971).

BLACK, L. P., GALE, N. H., MOORBATH, S., PANKHURST, R. J., McGREGOR, V. R.: Isotopic dating of very early Precambrian amphibolite facies gneisses from the Godthaab district, West Greenland. Earth Planet. Sci. Lett. **12**, 245—259 (1971).

BOELRIJK, N. A. I. M.: A general formula for double isotope dilution analysis. Chem. Geol. **3**, 323—325 (1968).

BOFINGER, V. M., COMPSTON, W.: A reassessment of the age of the Hamilton group, New York and Pennsylvania, and the role of inherited radiogenic Sr^{87}. Geochim. Cosmochim. Acta **31**, 2353—2359 (1967).

BOFINGER, V. M., COMPSTON, W., VERNON, M. J.: The application of acid leaching to Rb-Sr dating of a Middle Ordovician shale. Geochim. Cosmochim. Acta **32**, 823—834 (1968).

BOFINGER, V. M., COMPSTON, W., GULSON, B. L.: A Rb-Sr study of the Lower Silurian Stage Circle shale, Canberra, Australia. Geochim. Cosmochim. Acta **34**, 433—446 (1970).

BOGARD, D. D., BURNETT, D. S., EBERHARDT, P., WASSERBURG, G. J.: Rb^{87}-Sr^{87} isochron and K^{40}-Ar^{40} ages of the Norton County Chondrite. Earth Planet. Sci. Lett. **3**, 179—189 (1967).

BONATTI, E., HONNOREZ, J., FERRARA, G.: Equatorial Mid-Atlantic Ridge: Petrologic and Sr isotopic evidence for an alpine-type rock assemblage. Earth Planet. Sci. Lett. **9**, 247—256 (1970).

BONATTI, E.: Ancient continental mantle beneath oceanic ridges. J. Geophys. Res. **76**, 3825—3831 (1971).

BORCHERT, H., MUIR, R. O.: Salt deposits. London: Van Nostrand (1964).

BOTTINO, M. L., FULLAGAR, P. D.: The effects of weathering on whole-rock Rb-Sr ages of granitic rocks. Am. J. Sci. **266**, 661—670 (1968).

BOWEN, N. L.: The evolution of the igneous rocks. Reprinted by Dover Publications, New York 1956.

BRAITSCH, O.: Entstehung und Stoffbestand der Salzlagerstätten, Mineralogie und Petrographie in Einzeldarstellung. Berlin-Göttingen-Heidelberg: Springer 1962.

BRAITSCH, O.: Bromine and rubidium as indicators of environment during sylvite and carnallite deposition of the upper Rhine Valley evaporites, pp. 293—301. In: Symposium on Salt. Cleveland, Ohio: North Ohio Geol. Soc. Inc. 1966.

BROOKINS, D. G.: The strontium geochemistry of carbonates in kimberlites and limestones from Riley County, Kansas. Earth Plant. Sci. Lett. **2**, 235—240 (1967).

BROOKINS, D. G., CHAUDHURI, S., DOWLING, P. L.: The isotopic composition of strontium in Permian limestones, eastern Kansas. Chem. Geol. **4**, 439—444 (1969).

BROOKINS, D. G., WATSON, K. D.: The strontium geochemistry of calcite associated with kimberlite at Bachelor Lake, Quebec. J. Geol. **77**, 367—371 (1969).

BROOKS, C., COMPSTON, W.: The age and initial Sr^{87}/Sr^{86} of the Heemskirk granite, western Tasmania. J. Geophys. Res. **70**, 6249—6262 (1965).

BROOKS, C.: The effect of mineral age discordancies on total-rock Rb-Sr isochrons of the Heemskirk Granite, Western Tasmania. J. Geophys. Res. **71**, 5447—5458 (1966).

BROOKS, C.: Relationship between feldspar alteration and the precise post-crystallization movement of rubidium and strontium isotopes in a granite. J. Geophys. Res. **73**, 4751—4757 (1968).

BROOKS, C. K.: On the interpretation of trends in element ratios in differentiated igneous rocks, with particular reference to strontium and calcium. Chem. Geol. **3**, 15—20 (1968).

BROWN, P. E.: The Songwe scarp carbonatite and associated feldspathization in the Mbeya Range, Tanganyika. Quart. J. Geol. Soc. London **120**, 223—240 (1964).

BURCHART, J.: Rubidium-strontium isochron ages of the crystalline core of the Tatra Mountains, Poland. Am. J. Sci. **266**, 895—907 (1968).

BURNETT, D. S., WASSERBURG, G. J.: Evidence for the formation of an iron meteorite at 3.8×10^9 years. Earth Planet. Sci. Lett. **2**, 137—147 (1967a).

BURNETT, D. S., WASSERBURG, G. J.: Rb^{87}-Sr^{87} ages of silicate inclusions in iron meteorites. Earth Planet. Sci. Lett. **2**, 397—408 (1967b).

BURNETT, D. S.: Formation times of meteorites and lunar samples. EOS, Trans. Amer. Geophys. Union **52**, No. 7, 435—440 (1971).

BURST, J. F.: Glauconite pellets, their mineral nature and applications to stratigraphic interpretations. Am. Assoc. Petrol. Geol. Bull. **2**, 310—327 (1958).

CAHEN, L., DELHAL, J., DEUTSCH, S.: Rubidium-strontium geochronology of some granitic rocks from the Kibaran Belt (Central Katanga, Republic of the Congo). Musee Royal de L'Afrique Centrale, Tervuren, Belgique Annales, Ser. 1N-8°, Science Geologiques, No. 59 (1967).

CAHEN, L., DELHAL, J., DEUTSCH, S., GRÖGLER, N., LEDENT, D., PASTEELS, P.: Three contributions to the geochronology and petrogenesis of granitic rocks in the copperbelt of Zambia and southeast Katanga Province (Republic of the Congo) Musee Royal de L'Afrique Centrale, Tervuren, Belgique Annales, Ser. 1N-8°, Science Geologiques, No. 65 (1970).

CAMPBELL, N. R., WOOD, A.: The radioactivity of the alkali metals. Proc. Cambridge Phil. Soc. **14** S, 15—21 (1906).

CATANZARO, E. J., MURPHY, T. J., GARNER, E. L., SHIELDS, W. R.: Absolute isotopic abundance ratio and atomic weight of terrestrial rubidium. J. Res. Nat. Bur. Std. Physics and Chemistry **73** A, No. 5, 511—516 (1969).

CARSLAW, H. S., JAEGER, J. C.: Conduction of Heat in Solids. London: Oxford University Press 1959.

CHAUDHURI, S., FAURE, G.: Geochronology of the Keweenawan rocks, White Pine, Michigan. Econ. Geol. 62, 1011—1033 (1967).

CHAUDHURI, S., FAURE, G.: Rubidium-strontium age of the Mt. Bohemia intrusion in Michigan. J. Geol. 76, 488—490 (1968).

CHAUDHURI, S., BROOKINS, D. G.: The Rb-Sr whole-rock age of the Stearns shale (Lower Permian), eastern Kansas, before and after acid leaching experiments. Geol. Soc. Am. Bull. 80, 2605—2610 (1969).

CHAUDHURI, S., BROOKINS, D. G., FENTON, M. D.: Rubidium-strontium whole-rock and mineral ages of the Coldwell, Ontario, syenites. Abstracts with Programs, Geol. Soc. Am. 3, No. 4, 255 (1971).

CLOUD, P. E., JR.: Physical limits of glauconite formation. Am. Assoc. Petrol. Geol. Bull. 39, 484 (1955).

COMPSTON, W., JEFFERY, P. M.: Anomalous common strontium in granite. Nature 184, 1792—1793 (1959).

COMPSTON, W., JEFFERY, P. M., RILEY, G. H.: Age of emplacement of granites. Nature 186, 702—703 (1960).

COMPSTON, W., JEFFERY, P. M.: Metamorphic chronology by the rubidium-strontium method. Ann. N. Y. Acad. Sci. 91, Art. 2, 185—191 (1961).

COMPSTON, W., PIDGEON, R. T.: Rubidium-strontium dating of shales by the total-rock method. J. Geophys. Res. 67, 3493—3502 (1962).

COMPSTON, W., LOVERING, J. F., VERNON, M. J.: The rubidium-strontium age of the Bishopville aubrite and its component enstatite and feldspar. Geochim. Cosmochim. Acta 29, 1085—1099 (1965).

COMPSTON, W., McDOUGALL, I., HEIER, K. S.: Geochemical comparison of the Mesozoic basaltic rocks of Antarctica, South Africa, South America, and Tasmania. Geochim. Cosmochim. Acta 32, 129—150 (1968).

COMPSTON, W., LOVERING, J. F.: The strontium isotopic geochemistry of granulitic and eclogitic inclusions from the basic pipes at Delegate, eastern Australia. Geochim. Cosmochim. Acta 33, 691—699 (1969).

COMPSTON, W., ARRIENS, P. A., VERNON, M. J., CHAPPELL, B. W.: Rubidium-strontium chronology and chemistry of lunar material. Science 167, 474—476 (1970a).

COMPSTON, W., CHAPPELL, B. W., ARRIENS, P. A., VERNON, M. J.: The chemistry and age of Apollo 11 lunar material. Proc. Apollo 11 Lunar Science Conf. 2, 1007—1028 (1970b).

COOPER, J. A., RICHARDS, J. R.: Isotopic and alkali measurements from the Vema seamount of the south Atlantic Ocean. Nature 210, 1245—1246 (1966).

CORMIER, R. F.: Rubidium-strontium ages of the mineral glauconite and their application to the construction of a post-Precambrian time scale. Fourth Annual Progress Report for 1956/57, Dept. of Geology and Geophysics, M.I.T. (1957).

COTTON, A. B., WILKINSON, G.: Advanced inorganic chemistry. New York: Interscience Publishers, John Wiley 1962.

COX, J. M., FAURE, G.: Isotope composition of strontium in the carbonate phase of cores 1474P and 1445P from the Black Sea. In: The Black Sea; Its Chemistry, Geology and Biology. Am. Assoc. Petrol. Geol. Mem. (1972). In press.

COX, K. G., GASS, I. G., MALLICK, D. I. J.: The peralkaline volcanic suite of Aden and Little Aden, South Arabia, J. Petrol. 11, 433—462 (1970).

CRAMER, H. R.: Evaporites — a selected bibliography. Am. Assoc. Petrol. Geol. Bull. 53, 982—1011 (1969).

CUMMING, G. L.: A recalculation of the age of the solar system. Can. J. Earth Sci. 6, 719—736 (1969).

DALY, R. A.: Igneous rocks and the depths of the Earth. New York: McGraw-Hill 1933.

DALY, R. A.: Meteorites and an Earth-model. Geol. Soc. Am. Bull. 54, 401—456 (1943).

DASCH, E. J., HILLS, F. A., TUREKIAN, K. K.: Strontium isotopes in deep-sea sediments. Science 153, 295—297 (1966).

DASCH, E. J.: Strontium isotopes in weathering profiles, deep-sea sediments, and sedimentary rocks. Geochim. Cosmochim. Acta 33, 1521—1552 (1969).

DASCH, E. J.: Strontium isotope disequilibrium in a porphyritic alkali basalt and its bearing on magmatic processes. J. Geophys. Res. 74, 560—565 (1969).

DASCH, E. J., BISCAYE, P. E.: Isotopic composition of strontium in Cretaceous-to-Recent pelagic foraminifera. Earth Planet. Sci. Lett. 11, 201—204 (1971).

DAVIES, R. D., ALLSOPP, H. L., ERLANK, A. J., MANTON, W. I.: Sr-isotopic studies on various layered mafic intrusions in southern Africa. Geol. Soc. South Africa, Spec. Pub. 1. Symposium on the Bushveld Igneous Complex and Other Layered Intrusions 576—693 (1970).

DAWSON, J. B.: Geochemistry and the origin of kimberlite. In: WYLLIE, P. J.: Ultramafic and Related Rocks. New York: John Wiley 1967.

DEANS, T.: Isotopic composition of strontium in carbonatites. Nature 201, 599 (1964).

DEANS, T., SNELLING, N. J., RAPSON, J. E.: Strontium isotopes and trace elements in carbonatites and limestones from Ice River, British Columbia. Nature 210, 290—291 (1966).

DEANS, T., POWELL, J. L.: Trace elements and strontium isotopes in carbonatites, fluorites, and limestones from India and Pakistan. Nature 218, 750—752 (1968).

DEUSER, W. G., HERZOG, L. F.: Rubidium-strontium age determinations of muscovites and biotites from pegmatites of the Blue Ridge and Piedmont. J. Geophys. Res. 67, 1997—2004 (1962).

DEWEY, J. F., BIRD, J. M.: Mountain belts and the new global tectonics. J. Geophys. Res. 75, 2625—2647 (1970).

DICKINSON, D. R., DODSON, M. H., GASS, I. G., REX, D. C.: Correlation of initial Sr^{87}/Sr^{86} with Rb/Sr in some late Tertiary volcanic rocks of South Arabia. Earth Planet. Sci. Lett. 6, 84—90 (1969).

DICKINSON, W. R.: Relation of andesites, granites, and derivative sandstones to arc-trench tectonics. Rev. Geophys. Space Phys. 8, 813—860 (1970).

DOE, B. R., HART, S. R.: The effect of contact metamorphism on lead in potassium feldspars near the Eldora Stock, Colorado. J. Geophys. Res. 68, 3521—3530 (1963).

DOE, B. R., HEDGE, C. E., WHITE, D. E.: Preliminary investigation of the source of lead and strontium in deep geothermal brines underlying the Salton Sea geothermal area. Econ. Geol. 61, 462—483 (1966).

DOE, B. R.: Lead and strontium isotopic studies of Cenozoic volcanic rocks in the Rocky Mountain region — a summary. Quart. J. Colo. School Mines 63, 149—174 (1968).

DOE, B. R., LIPMAN, P. W., HEDGE, C. E., KURASAWA, H.: Primitive and contaminated basalts from the Southern Rocky Mountains, U.S.A., Contrib. Mineral. Petrol. 21, 142—156 (1968).

DOE, B. R., TILLING, R. I., HEDGE, C. E., KLEPPER, M. R.: Lead and strontium isotope studies of the Boulder Batholith, S. W. Montana. Econ. Geol. 63, 884—906 (1968).

DONN, W. L., DONN, B. V., VALENTINE, W. G.: On the early history of the Earth. Geol. Soc. Am. Bull. 76, 287—306 (1965).

DRAKE, M. J., McCALLUM, I. S., McKAY, G. A., WEILL, D. F.: Mineralogy and petrology of Apollo 12 sample No. 12013: A progress report. Earth Planet. Sci. Lett. 9, 103—123 (1970).

DUCKWORTH, H. E.: Mass Spectroscopy. Cambridge University Press 1960.

ECKERMANN, H. VON: The alkaline district of Alnö Island. Sveriges Geol. Undersok. Arsbok Ser. Ca, 36, 1—176 (1948).

ENGEL, A. E. J., ENGEL, C. G., HAVENS, R. G.: Chemical characteristics of oceanic basalts and the upper mantle. Geol. Soc. Am. Bull. 76, 719—734 (1965).

ENGEL, A. E. J., NAGY, B., NAGY, L. A., ENGEL, C. G., KREMP, G. O. W., DREW, C. M.: Alga-like forms in Onverwacht Series, South Africa: Oldest recognized lifelike forms on earth. Science 161, 1005—1008 (1968).

ERICKSON, R. L.: Stratigraphy and petrology of the Tascotal Mesa quadrangle, Texas. Geol. Soc. Am. Bull. 64, 1353—1386 (1953).

ERLANK, A. J.: The terrestrial abundance relationship between potassium and rubidium. In: AHRENS, L. H. (ed.): Origin and Distribution of the Elements. New York: Pergamon Press 1968.

EVERNDEN, J. F., CURTIS, G. H., KISTLER, R. W., OBRADOVICH, J.: Argon diffusion in glauconite, microcline, sanidine, leucite, and phlogopite. Am. J. Sci. 258, 583—604 (1960).

EWART, A., STIPP, J. J.: Petrogenesis of the volcanic rocks of the Central North Island, New Zealand, as indicated by a study of Sr^{87}/Sr^{86} ratios, and Sr, Rb, K, U, and Th abundances. Geochim. Cosmochim. Acta 32, 699—736 (1968).

FAIRBAIRN, H. W., HURLEY, P. M., PINSON, W. H., JR.: The relation of discordant Rb-Sr mineral and rock ages in an igneous rock to its time of crystallization and the time of subsequent Sr^{87}/Sr^{86} metamorphism. Geochim. Cosmochim. Acta 23, 135—144 (1961).

FAIRBAIRN, H. W., FAURE, G., PINSON, W. H., HURLEY, P. M., POWELL, J. L.: Initial ratio of Sr^{87}/Sr^{86}, whole-rock age, and discordant biotite in the Monteregian Igneous Province, Quebec. J. Geophys. Res. 68, 6515—6522 (1963).

FAIRBAIRN, H. W., HURLEY, P. M., PINSON, W. H.: Initial Sr^{87}/Sr^{86} and possible sources of granitic rocks in Southern British Columbia. J. Geophys. Res. 69, 4889—4893 (1964).

FAIRBAIRN, H. W., FAURE, G., PINSON, W. H., JR., HURLEY, P. M.: Rb-Sr whole-rock age of the Sudbury lopolith and basin sediments. Can. J. Earth Sci. 5, pt. 2, 707—714 (1967).

FAIRBAIRN, H. W., HURLEY, P. M., CARD, K. D., KNIGHT, C. J.: Correlation of radiometric ages of Nipissing diabase and Huronian metasediments with Proterozoic orogenic events in Ontario. Can. J. Earth Sci. 6, 489—497 (1969).

FAURE, G.: The Sr^{87}/Sr^{86} ratio in oceanic and continental basalts and the origin of igneous rocks. Ph. D. dissertation, Dept. Geol. Geophys., Massachusetts Institute of Technology (1961).

FAURE, G., HURLEY, P. M.: The ratio Sr^{87}/Sr^{86} in oceanic and continental basalts. Abstract. J. Geophys. Res. 66, 2527 (1961).

FAURE, G., HURLEY, P. M., FAIRBAIRN, H. W., PINSON, W. H.: Isotopic compositions of strontium in continental basic intrusives. Abstract. J. Geophys. Res. 67, 3557 (1962).

FAURE, G., HURLEY, P. M.: The isotopic composition of strontium in oceanic and continental basalt: Application to the origin of igneous rocks. J. Petrol. 4, 31—50 (1963).

FAURE, G., HURLEY, P. M., FAIRBAIRN, H. W.: An estimate of the isotopic composition of strontium in rocks of the Precambrian Shield of North America. J. Geophys. Res. 68, 2323—2329 (1963).

FAURE, G., FAIRBAIRN, H. W., HURLEY, P. M., PINSON, W. H., JR.: Whole-rock Rb-Sr age of norite and micropegmatite at Sudbury, Ontario. J. Geol. 72, 848—854 (1964).

FAURE, G., HURLEY, P. M., POWELL, J. L.: The isotopic composition of strontium in surface water from the North Atlantic Ocean. Geochim. Cosmochim. Acta 29, 209—220 (1965).

FAURE, G., CROCKET, J. H., HURLEY, P. M.: Some aspects of the geochemistry of strontium and calcium in the Hudson Bay and the Great Lakes: Geochim. Cosmochim. Acta 31, 451—461 (1967).

FAURE, G., HILL, R. L., EASTIN, R., MONTIGNY, R. J. E.: Age determination of rocks and minerals from the Transantarctic Mountains. Antarctic J. U.S. 3, No. 5, 173—175 (1968a).

FAURE, G., MURTAUGH, J. G., MONTIGNY, R. J. E.: The geology and geochronology of the basement complex of the central Transantarctic Mountains. Can. J. Earth Sci. 5, 555—560 (1968b).

FAURE, G., CHAUDHURI, S., FENTON, M. D.: Age of the Duluth Gabbro and of the Endion Sill, Duluth, Minnesota. J. Geophys. Res. 74, 720—725 (1969).

FAURE, G., KOVACH, J.: The age of the Gunflint Formation of the Animikie Series in Ontario, Ontario, Canada. Geol. Soc. Am. Bull. 80, 1725—1736 (1969).

FAURE, G., JONES, L. M.: Anomalous strontium in the Red Sea brines. In: DEGENS, E. T., ROSS, D. A. (eds.): Hot Brines and Recent Heavy Metal Deposits in the Red Sea. Berlin-Heidelberg-New York: Springer 1969.

FAURE, G., HILL, R. L., JONES, L. M., ELLIOT, D. H.: Isotope composition of strontium and silica content of Mesozoic basalt and dolerite from Antarctica. SCAR Symposium. Oslo: University Press 1971.

FECHTIG, H., KALBITZER, S.: The diffusion of argon in potassium-bearing solids, pp. 68—106. In: SCHAEFFER, O. A., ZÄHRINGER, J. (eds.): Potassium Argon Dating. Berlin-Heidelberg-New York: Springer 1966.

FENTON, M. D., FAURE, G.: The isotopic evolution of terrestrial strontium. Abstract. Geol. Soc. Am. Program, Annual Meeting, 1, No. 7, p. 64 (1969).

FENTON, M. D., FAURE, G.: The age of the igneous rocks of the Stillwater Complex of Montana. Geol. Soc. Am. Bull. 80, 1599—1604 (1969).

FERRARA, G., CLARKE, W. B., MURTHY, V. R., BASS, M. N.: K-Ar ages of Juan Fernanadez Islands and Southeast Pacific dredge hauls. Abstract. Trans. Am. Geophys. Union 50, 329 (1969).

FISH, R. A., GOLES, G. G., ANDERS, E.: The record in the meteorites, III. On the development of meteorites in asteroidal bodies. Astrophys. J. 132, 243—258 (1960).

FLYNN, K. F., GLENDENIN, L. E.: Half-life-and beta spectrum of Rb^{87}. Phys. Rev. 116, 744—748 (1959).

FRIEDMAN, I.: Mass spectrometry. In: FAUL, H. (ed.): Nuclear Geology, pp. 64—70. New York: Wiley 1954.

FULLAGAR, P. D., LEMMON, R. E., RAGLAND, P. C.: Petrochemical and geochronological studies of putonic rocks in the Southern Appalachians: Part I. The Salisbury Pluton. Geol. Soc. Am. Bull. 82, 409—416 (1971).

GAST, P. W.: Abundance of Sr^{87} during geologic time. Geol. Soc. Am. Bull. 66, 1449—1454 (1955).

GAST, P. W.: Limitations on the composition of the upper mantle. J. Geophys. Res. 65, 1287—1297 (1960).

GAST, P. W.: The rubidium-strontium method. Ann. N. Y. Acad. Sci. 91, 181—184 (1961).

GAST, P. W.: The isotopic composition of strontium and the age of stone meterorites, I. Geochim. Cosmochim. Acta 26, 927—943 (1962).

GAST, P. W., TILTON, G. R., HEDGE, C. E.: Isotopic composition of lead and strontium from Ascension and Gough Islands. Science 145, 1181—1185 (1964).

GAST, P. W.: Isotope geochemistry of volcanic rocks. In: HESS, H. H., POLDERVAARTA. (eds.): Basalts: The Poldervaart Treatise on Rocks of Basaltic Composition, Vol. 1, pp. 325—358. New York: Interscience 1967.

GAST, P. W.: Trace element fractionation and the origin of tholeiitic and alkaline magma types. Geochim. Cosmochim. Acta 32, 1057—1086 (1968).

GAST, P. W., HUBBARD, N. J.: Abundance of alkali metals, alkaline and rare earths, and strontium-87/strontium-86 ratios in lunar samples. Science 167, 485—487 (1970).

GAST, P. W., HUBBARD, N. J., WIESMAN, H.: Chemical composition and petrogenesis of basalts from Transquillity Base. Proc. Apollo 11 Lunar Science Conf. 2, 1143—1164 (1970).

GEESE-BÄHNISCH, I., HUSTER, E.: Neubestimmung der Halbwertszeit des Rb[87]. Naturwissenschaften 41, 495 (1954).

GENTNER, W., PRÄG, R., SMITS, F.: Das Alter eines Kalilagers im unteren Oligozän. Geochim. Cosmochim. Acta 4, 11—20 (1953).

GENTNER, W., GOEBEL, K., PRÄG, R.: Argonbestimmungen und Kalium-Mineralien, III. Vergleichende Messungen nach der Kalium-Argon- und Uran-Helium-Methode. Geochim. Cosmochim. Acta 5, 124—133 (1954).

GERLING, E. K., SHUKOLYUKOV, Y. A.: Determination of the absolute age from the ratio of isotopes Sr^{87}/Sr^{86} in sedimentary rocks. Geochemistry 3, 226—230 (1957).

GERLING, E. K., YASCHENKO, M. L., VARSHAVSKAYA, E. S.: Relation between initial isotopic composition of strontium and age of granite rocks. Geochem. Intern. 5, 50—55 (1968).

GILETTI, B. J., MOORBATH, S., LAMBERT, R., ST. J.: A geochronological study of the metamorphic complexes of the Scottish Highlands. Quart. J. Geol. Soc. London 117, 233—272 (1961).

GILL, J. B.: Geochemistry of Viti Levu, Fiji, and its evolution as an island arc. Contrib. Mineral. Petrol. 27, 179—203 (1970).

GILLULY, J. (ed.): Origin of granite. Geol. Soc. Am. Mem. 28, 139 (1948).

GITTINS, J., HAYATSU, A., YORK, D.: A strontium isotope study of metamorphosed limestones. Lithos 3, 51—58 (1969).

GOLDSCHMIDT, V. M., BERMAN, H., HAUPTMAN, H., PETERS, O.: Zur Geochemie der Alkalimetalle I. Nachr. Akad. Ges. Wiss. Göttingen, Math.-Physik. Kl. IV, N. F. l. (1933).

GOLDSCHMIDT, V. M., BAUER, H., WITTE, H.: Zur Geochemie der Alkalimetalle II. Nachr. Akad. Wiss. Göttingen, Math.-Physik. Kl. IV, N. F. l., No. 4 (1934).

GOLDSCHMIDT, V. M.: Geochemische Verteilungsgesetze der Elemente. IX. Die Mengenverhältnisse der Elemente und der Atom-Arten: Skrifter Norske Videnskaps-Akad. Oslo I: Mat.-Naturv. Kl. IV, No. 4 (1937).

GOLDSCHMIDT, V. M.: Geochemistry. Oxford: Clarendon Press 1954.

GOLDICH, S. S., ELMS, M. A.: Stratigraphy and petrology of the Buck Hill Quadrangle, Texas. Geol. Soc. Am. Bull. 60, 1133—1182 (1949).

GOLDICH, S. S., GAST, P. W.: Effects of weathering on the Rb-Sr and K-Ar ages of biotite from the Morton Gneiss, Minnesota. Earth Planet. Sci. Letters 1, 372—375 (1966).

GOLDBERG, E. D.: Minor elements in sea water. In: RILEY, J. P., SKIRROW, G. (eds.): Chemical Oceanograph, Vol. 1, pp. 162—196. New York: Academic Press 1965.

GOLES, G. C., FISH, R. A., ANDERS, E.: The record in the meteorites I. The former environment of stone meteorites as deduced from K^{40}-Ar^{40} ages. Geochim. Cosmochim. Acta **19**, 177—196 (1960).

GOPALAN, K., WETHERILL, G. W.: Rubidium-strontium age of hypersthene (L) chondrites. J. Geophys. Res. **73**, 7133—7136 (1968).

GOPALAN, K., WETHERILL, G. W.: Rubidium-strontium age of amphoterite (LL) chondrites. J. Geophys. Res. **74**, 4349—4358 (1969).

GOPALAN, K., KAUSHAL, S., LEE-HU, C., WETHERILL, G. W.: Rubidium-strontium, uranium, and thorium-lead dating of lunar material. Science **167**, 471—473 (1970a).

GOPALAN, K., KAUSHAL, S., LEE-HU, C., WETHERILL, G. W.: Rb-Sr and U, Th-Pb ages of lunar materials. Proc. Apollo 11 Lunar Science Conf. **2**, 1143—1164 (1970b).

GOPALAN, K., WETHERILL, G. W.: Rubidium-strontium studies on enstatite chondrites: Whole meteorite and mineral isochrons. J. Geophys. Res. **75**, 3457—3467 (1970).

GRANT, J. A.: Rubidium-strontium isochron study of the Grenville front near Lake Timagami, Ontario. Science **146**, 1049—1053 (1964).

GREEN, D. H., RINGWOOD, A. E.: The genesis of basaltic magmas. Contrib. Mineral. Petrol. **15**, 103—190 (1967).

GREEN, J.: Geochemical table of the elements for 1959. Geol. Soc. Am. Bull. **70**, 1127—1184 (1959).

GRIFFIN, W. L., MURTHY, V. R.: Abundances of K, Rb, Sr, and Ba in some ultrabasic rocks and minerals. Earth Planet. Sci. Letters **4**, 497—501 (1968).

GRIFFIN, W. L., MURTHY, V. R.: Distribution of K, Rb, Sr, and Ba in some minerals relevant to basalt genesis. Geochim. Cosmochim. Acta **33**, 1389—1414 (1969).

GRIM, R. E.: Clay Mineralogy. New York: McGraw-Hill 1952.

GULBRANDSEN, R. A., GOLDICH, S. S., THOMAS, H. H.: Glauconite from the Precambrian Belt Series, Montana. Science **140**, 390—391 (1963).

HAHN, O., STRASSMAN, F., WALLING, E.: Herstellung wägbarer Mengen des Strontiumisotops 87 als Umwandlungsprodukt des Rubidiums aus einem kanadischen Glimmer. Naturwissenschaften **25**, 189 (1937).

HAHN, O., WALLING, E.: Über die Möglichkeit geologischer Altersbestimmungen rubidiumhaltiger Mineralen und Gesteine. Z. Anorg. Allgem. Chem. **236**, 78—82 (1938).

HAHN, O., STRASSMAN, F., MATTAUCH, J., EWALD, H.: Geologische Altersbestimmungen mit der Strontiummethode. Chem. Zeitung **67**, 55—56 (1943).

HALPERN, M.: Sr^{87}/Sr^{86} ratios of ultramafic nodules and host basalt from the McMurdo area and Ford Ranges, Antarctica. Antarctic J. U.S. **4**, 206 (1969).

HAMILTON, E. I., DEANS, T.: Isotopic composition of strontium in some African carbonatites and limestones and in strontium minerals. Nature **198**, 776—777 (1963).

HAMILTON, E. I.: The isotopic composition of strontium in the Skaergaard Intrusion, East Greenland. J. Petrol. **4**, 383—391 (1964).

HAMILTON, E. I.: Isotopic composition of strontium in carbonatites. Nature **198**, 776—777 (1964).

HAMILTON, E. I.: Distribution of some trace elements and the isotopic composition of strontium in Hawaiian lavas. Nature **206**, 251—253 (1965).

HAMILTON, E. I.: Applied Geochronology. New York: Academic Press 1965.

HAMILTON, E. I.: Isotopic composition of strontium in a variety of rocks from Reunion Island. Nature **207**, 1188 (1965).

HAMILTON, E. I.: The isotopic composition of strontium in Atlantic Ocean water. Earth Planet. Sci. Letters. **1**, 435—436 (1966).

HAMILTON, E. I.: The isotopic composition of strontium applied to problems of the origin of the alkaline rocks. In: HAMILTON, E. I., FARQUHAR, R. M. (eds.): Radiometric Dating for Geologists, pp. 437—463. New York: Interscience 1968.

HAMILTON, W., MEYERS, W. B.: The nature of batholiths. U.S. Geol. Surv. Prof. Paper 554-C (1967).

HAMILTON, W.: Mesozoic California and the underflow of Pacific mantle. Geol. Soc. Am. Bull. **80**, 2409—2430 (1969a).

HAMILTON, W.: The volcanic central Andes — a modern model for the Cretaceous batholiths and tectonics of western North America. Oregon, Dep. Geol. Mineral Ind. Bull. **65**, 175—184 (1969b).

HANSON, G. N., GAST, P. W.: Kinetic studies in contact metamorphic zones. Geochim. Cosmochim. Acta **31**, 1119—1153 (1967).

HANSON, G. N., GRÜNENFELDER, M., SOPTRAYANOVA, G.: The geochronology of recrystallized tectonite in Switzerland — The Roffna Gneiss. Earth Planet. Sci. Letters **5**, 413—422 (1969).

HARKER, A.: The Tertiary Igneous Rocks of Skye. Glasgow: J. Hedderwick and Sons 1904.

HART, S. R.: The petrology and isotopic mineral age relations of a contact zone in the Front Range, Colorado. J. Geol. **72**, 493—525 (1964).

HART, S. R., TILTON, G. R.: The isotope geochemistry of strontium and lead in Lake Superior sediments and water. In: STEINHART, J. S., SMITH, T. L., (eds.): The Earth beneath the Continents, Geophysical Monograph No. 10. Am. Geophys. Union 127—137 (1966).

HART, S. R., DAVIS, G. L., STEIGER, R. H., TILTON, G. R.: A comparison of the isotopic mineral age variations and petrologic changes induced by contact metamorphism. In: HAMILTON, E. I., FARQUHAR, R. M. (eds.): Radiometric Dating for Geologists, pp. 73—110. New York: Interscience Publ. 1968.

HART, S. R.: Isotope geochemistry of crust-mantle processes. In: HART, P. J. (ed.): The Earth's Crust and Upper Mantle, Geophysical Monograph 13. Am. Geophys. Union 58—62 (1969)

HART, S. R., BROOKS, C.: Rb-Sr mantle evolution models. Carnegie Inst. Wash. Yearbook **69**, 426—429 (1970).

HART, S. R., NALWALK, A. J.: K, Rb, Cs and Sr relationships in submarine basalts from the Puerto Rico trench. Geochim. Cosmochim. Acta **34**, 145—155 (1970).

HAYATSU, A., FARQUHAR, R. M., YORK, D., GITTINS, J.: Precise measurement of Sr isotope ratios and their variation in metamorphosed limestone. Abstract. Trans. Amer. Geophys. Union **46**, 174 (1965).

HAYATSU, A., YORK, D., FARQUHAR, R. M., GITTINS, J.: Significance of strontium isotope ratios in theories of carbonatite genesis. Nature **207**, 625—626 (1965).

HEATH, S. A.: Sr^{87}/Sr^{86} ratios in anorthosites and some associated rocks. Fifteenth Ann. Prog. Rep., M.I.T. 65—88 (1967).

HEDGE, C. E., WALTHALL, F. G.: Radiogenic strontium-87 as an index of geological processes. Science **140**, 1214—1217 (1963).

HEDGE, C. E.: Significance of radiogenic strontium in volcanic lavas. Abstract. Trans. Am. Geophys. Union **45**, 114 (1964).

HEDGE, C. E.: Variations in radiogenic strontium found in volcanic rocks. J. Geophys. Res. **71**, 6119—6126 (1966).

HEDGE, C. E., PETERMAN, Z. E.: Sr^{87}/Sr^{86} of circum-Pacific andesites. Geol. Soc. Am., Abstracts with Programs **1**, Pt. 7, 96 (1969).

HEDGE, C. E., PETERMAN, Z. E.: The strontium isotopic composition of basalts from the Gorda and Juan de Fuca Rises, Northeastern Pacific Ocean. Contrib. Min. Petrol. **27**, 114—120 (1970).

HEDGE, C. E., HILDRETH, R. A., HENDERSON, W. T.: Strontium isotopes in some Cenozoic lavas from Oregon and Washington. Earth Planet Sci. Letters **8**, 434—438 (1970).

HEIER, K. S., TAYLOR, S. R.: The distribution of Li, Na, K, Rb, Cs, Pb, and Tl in Southern Norwegian pre-Cambrian feldspars. Geochim. Cosmochim. Acta **15**, 284—304 (1959).

HEIER, K. S.: Rubidium/strontium and Sr^{87}/Sr^{86} ratios in deep crustal material. Nature **202**, 477—478 (1964).

HEIER, K. S., ADAMS, J. A. S.: The geochemistry of the alkali metals. Phys. Chem. Earth **5**, 253—281 (1964).

HEIER, K. S., COMPSTON, W., McDOUGALL, I.: Thorium and uranium concentrations, and the isotopic composition of strontium in the differentiated Tasmanian dolerites. Geochim. Cosmochim. Acta **29**, 643—659 (1965).

HEIER, K. S., CHAPPELL, B. W., ARRIENS, P. A., MORGAN, J. W.: The geochemistry of four Icelandic basalts. Norsk Geol. Tidsskr. **46**, 427—437 (1966).

HEIER, K. S., COMPSTON, W.: Interpretation of Rb-Sr age patterns in high-grade metamorphic rocks, North Norway. Norsk Geol. Tidsskr. **49**, 257—284 (1969 a).

HEIER, K. S., COMPSTON, W.: Rb-Sr isotopic studies of the plutonic rocks of the Oslo Region. Lithos **2**, 133—145 (1969 b).

HEINRICH, E. W.: The geology of carbonatites. Chicago: Rand-McNally 1966.

HEMMENDINGER, A., SMYTHE, W. R.: Radioactive isotope of rubidium: Phys. Rev. **51**, 1052—1053 (1937).

HERZ, N.: Anorthosite belts, continental drift, and the anorthosite event. Science **164**, 944—947 (1969).

HERZOG, L. F., ALDRICH, L. T., HOLYK, W. K., WHITING, F. B., AHRENS, L. H.: Variations in strontium isotope abundances in minerals. Part 2: Radiogenic Sr^{87} in biotite, feldspar and celestite. Am. Geophs. Union Trans. **34**, 461—470 (1953).

HERZOG, L. F.: Rb and Sr and Sr isotopic content of the Homestead and Forest City chondrites. Astron. J. **60**, 163 (1955).

HERZOG, L. F., PINSON, W. H.: Rb/Sr age, elemental and isotopic abundance studies of stony meteorites. Am. J. Sci. **254**, 555 (1956).

HERZOG, L. F., PINSON, W. H., CORMIER, R. F.: Sediment age determination by Rb/Sr analysis of glauconite. Am. Ass. Petrol. Geol. Bull. **42**, 717—733 (1958).

HEVESY, G. V., WÜRSTLIN, K.: Über die Häufigkeit des Strontiums. Z. Anorg. Allgem. Chem. **216**, 312 (1934).

HILDRETH, R. A., HENDERSON, W. T.: Comparison of Sr^{87}/Sr^{86} for sea-water strontium and the Eimer and Amend $SrCO_3$. Geochim. Cosmochim. Acta 35, 235—238 (1971).

HOEFS, J., WEDEPOHL, K. H.: Strontium isotope studies on young volcanic rocks from Germany and Italy. Contrib. Mineral. Petrol. **19**, 328—338 (1968).

HOLMES, A.: The origin of igneous rocks. Geol. Mag. Lond. **69**, 550—558 (1932).

HOLMES, A.: The construction of a geologic time scale. Trans. Geol. Soc. Glasgow **21**, Pt. I, 117 (1947).

HOLMES, A.: Petrogenesis of Ratungite and its associates. Am. Mineral. **35**, 772—792 (1950).

HORSTMAN, E. L.: The distribution of lithium, rubidium, and caesium in igneous and sedimentary rocks. Geochim. Cosmochim. Acta **12**, 1—28 (1957).

HUBBARD, N. J., GAST, P. W., MEYER, C.: The origin of the lunar soil based on REE, K, Rb, Ba, Sr, P, and Sr^{87}/Sr^{86} data. Abstract. 1971 Lunar Science Conference Abstracts, January 11—14, 1971, NASA, Houston, Texas, p. 246 (1971).

HURLEY, P. M., CORMIER, R. F., HOWER, J., FAIRBAIRN, H. W., PINSON, W. H., JR.: Reliability of glauconite for age measurement by K-Ar and Rb-Sr methods. Am. Assoc. Petrol. Geol. Bull. **44**, 1793 (1960).

HURLEY, P. M., HUGHES, H., FAURE, G., FAIRBAIRN, H. W., PINSON, W. H.: Radiogenic strontium-87 model of continent formation. J. Geophys. Res. **67**, 5315—5334 (1962).

HURLEY, P. M., FAURE, G., HUGHES, H., FAIRBAIRN, H. W., PINSON, W. H., JR.: Evidence of continuing separation of sial from the mantle from the isotopic composition of common strontium. In: Nuclear geophysics, Nuclear Sci. Ser., Rept. No. 38, Nat. Acad. Sci.-Nat. Res. Council, Publ. 1075, 83—92 (1963).

HURLEY, P. M., FAIRBAIRN, H. W., FAURE, G., PINSON, W. H., Jr.: New approaches to geochronology by strontium isotope variations in whole rocks. In: Radioactive Dating Internat. At. Energy Agency, 201—218 (1963).

HURLEY, P. M., BATEMAN, P. C., FAIRBAIRN, H. W., PINSON, W. H., Jr.: Investigation of initial Sr^{87}/Sr^{86} ratios in the Sierra Nevada Plutonic Province. Geol. Soc. Am. Bull. **76**, 165—174 (1965).

HURLEY, P. M., FAIRBAIRN, H. W., PINSON, W. H., JR.: Rb-Sr isotopic evidence in the origin of potash-rich lavas of western Italy. Earth Planet. Sci. Letters **5**, 301—306 (1966).

HURLEY, P. M.: K-Ar dating of sediments. In: SCHAEFFER, O. A., ZÄHRINGER, J. (eds.): Potassium Argon Dating. Berlin-Heidelberg-New York: Springer 1966.

HURLEY, P. M.: $Rb^{87}-Sr^{87}$ relationships in the differentiation of the mantle. In: WYLLIE, P. J. (ed.): Ultramafic and Related Rocks. New York: John Wiley 1967.

HURLEY, P. M.: Absolute abundance of Rb, K, and Sr in the earth. Geochim. Cosmochim. Acta **32**, 273—284 (1968a).

HURLEY, P. M.: Correction to: Absolute abundance and distribution of Rb, K, and Sr in the earth. Geochim. Cosmochim. Acta **32**, 1025—1030 (1968b).

HURLEY, P. M., RAND, J. R.: Pre-drift continental nuclei. Science **164**, 1229—1242 (1969).

HURLEY, P. M., PINSON, W. H., JR.: Rubidium-strontium relations in Tranquillity Base samples. Science **167**, 473—474 (1970a).

HURLEY, P. M., PINSON, W. H., JR.: Whole-rock Rb-Sr isotopic age relationships in Apollo 11 lunar samples. Proc. Apollo 11 Lunar Sci. Conf. **2**, 1311—1316 (1970b).

HUTCHISON, R., DAWSON, J. B.: Rb, Sr, and $^{87}Sr/^{86}Sr$ in ultrabasic xenoliths and host rocks, Lashaine Volcano, Tanzania. Earth Planet. Sci. Letters **9**, 87—92 (1970).

IKPEAMA, M. O. U.: Strontium isotope composition of sediment and fossil shells from the Discovery Deep, Red Sea. M. Sc. thesis, Department of Geology. The Ohio State University, Columbus, Ohio 1971.

INGHRAM, M. G.: Modern mass spectroscopy. In: Advances in Electronics, Vol. 1, pp. 219—268. New York: Academic Press 1948.

JÄGER, E.: Rb-Sr age determinations on micas and total rocks from the Alps. J. Geophys. Res. **67**, 5293—5306 (1962).

JÄGER, E., ZWART, H. J.: Rb-Sr age determinations of some gneisses and granites of the Aston-Hospitalet Massif (Pyrenees). Geol. Mijnbouw **47**, 349—357 (1968).

JAEGER, J. C.: The temperature in the neighborhood of a cooling intrusive sheet. Am. J. Sci. **255**, 306—318 (1957).

JONES, L. M., FAURE, G.: Origin of the salts in Lake Vanda, Wright Valley, Southern Victoria Land, Antarctica. Earth Planet. Sci. Letters **3**, 101—106 (1967).

JONES, L. M., FAURE, G., MONTIGNY, R. J. E.: Geochemical studies in Wright Valley. Antarctic J. U.S. **2**, 114 (1967).

KAUSHAL, S. K., WETHERILL, G. W.: Rb^{87}-Sr^{87} age of bronzite (H group) chondrites. J. Geophys. Res. **74**, 2717—2726 (1969).

KAUSHAL, S. K., WETHERILL, G. W.: Rubidium 87-strontium 87 age of carbonaceous chondrites. J. Geophys. Res. **75**, 463—468 (1970).

KAZAKOV, G. A.: The use of glauconite to determine the absolute age of sedimentary rocks. Chemiya Zemm. Kory. **2**, Moscow (Nauka), 539—551 (1964).

KEMPE, W., MÜLLER, O.: The stony meteorite Krähenberg: Its chemical composition and the Rb-Sr age of the light and dark portions. In: MILLMAN, P. M. (ed.): Meteorite Research, pp. 418—428. Dortrecht, Holland: Reidel Publ. Co. 1969.

KINSMAN, D. J. J.: Interpretation of Sr^{+2} concentrations in carbonate minerals and rocks. J. Sediment Petrol. **39**, 486—508 (1969).

KOVACH, A.: A redetermination of the half-life of rubidium-87. Acta Phys. Acad. Sci. Hung. **17**, 341—351 (1966).

KOVACH, A., BALOGH, K., PANTO, G.: Strontium isotopic ratios in Tertiary igneous rocks of the Tokaj Mountains, northeastern Hungary. Acta Geol. (Budapest) **12**, 79—97 (1968).

KOVACH, A.: Strontium isotopes in some Tertiary igneous rocks in Hungary. Abstract. Symposium on volcanoes and their roots, Oxford, IAVCEI, 131—132 (1969).

KRANKOWSKY, D., ZÄHRINGER, J.: K-Ar ages of meteorites. In: SCHAEFFER, O. A., ZÄHRINGER, J. (ed.): Potassium Argon Dating, pp. 174—200. Berlin-Heidelberg-New York: Springer 1966.

KRAUSKOPF, K. B.: Introduction to Geochemistry. New York: McGraw-Hill 1967.

KRINOV, E. L.: Principles of Meteoritics. New York: Pergamon Press 1960.

KROGH, T. E., HURLEY, P. M.: Strontium isotope variation and whole-rock isochron studies, Grenville Province, Ontario. J. Geophys. Res. **73**, 7107—7125 (1968).

KROGH, T. E., DAVIS, G. L.: Old isotopic ages in the northwestern Grenville Province, Ontario. Geol. Ass. Can. Spec. Pap. **5**, 189—192 (1969).

KRUMBEIN, W. C.: Occurrence and lithologic associations of evaporites in the United States. J. Sediment Petrol **21**, 63—81 (1951).

KULP, J. L., TUREKIAN, K. K., BOYD, D. W.: Strontium content of limestones and fossils: Geol. Soc. Am. Bull. **63**, 701—716 (1952).

KULP, J. L., ENGELS, J.: Discordances in K-Ar and Rb-Sr isotopic ages. In: Radioactive Dating, pp. 219—238. Intern. At. Energy Agency 1963.

KULP, J. L.: Concordance and discordance between K-A and Rb-Sr isotopic ages from micas. In: VINOGRADOV, A. P. (ed.): Chemistry of the Earth's Crust, Vol. II, pp. 592—606. Israel Program for Scientific Translations Jerusalem 1967.

KURASAWA, H.: Lead and strontium isotopes of volcanic rocks of Japan. Abstract. Symposium on volcanoes and their roots, Oxford. IAVCEI, 133—134 (1969).

LAMBERT, D. L., WARNER, B.: The abundance of the elements in the solar photo-sphere-V, The alkaline earths, Mg, Ca, Sr, Ba. Monthly Notices-Roy. Astron. Soc. **140**, 197 (1968).

LAMBERT, D. L., MALLIA, E. A.: The abundances of the elements in the solar photosphere-VI, Rubidium. Monthly Notices Roy. Astron. Soc. **140**, 13 (1968).

LANPHERE, M. A., WASSERBURG, G. J. F., ALBEE, A. L., TILTON, G. R.: Redistribu-tion of strontium and rubidium isotopes during metamorphism, World Beater Complex, Panamint Range, California. In: CRAIG, H., MILLER, S. L., WASSERBURG, G. J. (eds.): Isotope and Cosmic Chemistry, pp. 269—320. Amsterdam: North-Holland Publ. Co. 1964.

LANPHERE, M. A.: Sr-Rb-K and Sr isotopic relationships in ultramafic rocks, southeastern Alaska. Earth Planet. Sci. Letters **4**, 185—190 (1968).

LAUGHLIN, A. W., BROOKINS, D. G., KUDO, A. M., CAUSEY, J. D.: Chemical and strontium isotopic investigations of ultramafic inclusions and basalt, Bandera Crater, New Mexico. Geochim. Cosmochim. Acta **35**, 107—113 (1971).

LEEMAN, W. P.: The isotopic composition of strontium in late-Cenozoic basalts from the Basin-Range Province, western United States. Geochim. Cosmochim. Acta **34**, 857—872 (1970).

LEGGO, P. J., HUTCHISON, R.: A Rb-Sr isotope study of ultrabasic xenoliths and their basaltic host rocks from the Massif Central, France. Earth Planet. Sci. Letters **5**, 71—75 (1968).

LESSING, P., DECKER, R. W., REYNOLDS, R. P., JR.: Potassium and rubidium distribution in Hawaiian lavas. J. Geophys. Res. **68**, 5851—5855 (1963).

LESSING, P., CATANZARO, E. J.: Sr^{87}/Sr^{86} ratios in Hawaiian lavas. J. Geophys. Res. **69**, 1599—1601 (1964).

LEUTZ, H., WENNINGER, H., ZIEGLER, K.: Die Halbwertszeit des Rb^{87}. Z. Physik **169**, 409—416 (1962).

LEWIS, J. F.: Trace elements, variation in alkalis, and the ratio Sr^{87}/Sr^{86} in selected rocks from the Taupo Volcanic zone. New Zealand J. Geol. Geophys. **11**, 608—629 (1968).

LONG, L. E., LAMBERT, R. ST. J.: Rb-Sr isotopic ages from the Moine series. In: JOHNSON, M. R. W., STEWART, F. H. (eds.): The British Caledonides, pp. 217—247. Edinburgh: Oliver and Boyd, Ltd. 1963.

LONG, L. E.: Rb-Sr chronology of the Carn Chuinneag Intrusion, Ross-shire, Scotland. J. Geophys. Res. **69**, 1589—1597 (1964).

LONG, L. E.: Isotope dilution analyses of common and radiogenic strontium using Sr^{84}-enriched spike. Earth Planet. Sci. Letters **1**, 289—292 (1966).

LOTZE, F.: Steinsalz und Kalisalze Geologie. Berlin: Gebr. Bornträger 1938.

LOVERING, J. F.: Pressures and temperatures within a typical parent meteorite body. Geochim. Cosmochim. Acta **12**, 253—261 (1957).

LOVERING, J. F.: A typical parent meteorite body. Geochim. Cosmochim. Acta **14**, 174—177 (1958).

LOVERING, T. S.: Theory of heat conduction applied to geological problems. Geol. Soc. Am. Bull. **46**, 69—94 (1935).

LOWENSTAM, H. A.: Sr/Ca ratio of skeletal aragonites from the Recent marine biota at Palau and from fossil gastropods. In: CRAIG, H., MILLER, S. L., WASSERBURG, G. J. (eds.): Isotopic and Cosmic Chemistry, pp. 114—132. Amster-dam: North-Holland Publ. Co. 1964.

Lunatic Asylum: Mineralogical and isotopic investigations on lunar rock 12013. Earth Planet. Sci. Letters **9**, 137—163 (1970).

Lunatic Asylum: Rb-Sr ages, chemical abundance patterns and history of lunar rocks. Abstract. 1971 Lunar Sci. Conf., January 11—14, 1971, NASA, Houston, Texas, 56—57 (1971).

MACGREGOR, M. H., WIEDENBECK, M. L.: The third forbidden beta spectrum of rubidium-87. Phys. Rev. **94**, 138 (1954).

MANTON, W. I.: The origin of associated basic and acid rocks in the Lebombo-Nuanetsi igneous province, Southern Africa, as implied by strontium isotopes. J. Petrol. **9**, 23—39 (1968).

MANTON, W. I., TATSUMOTO, M.: Some Pb and Sr isotopic measurements on eclogites from the Roberts Victor Mine, South Africa. Earth Planet. Sci. Letters **10**, 217—226 (1971).

MASON, B.: Principles of Geochemistry. New York: John Wiley 1958.

MASON, B.: Meteorites. New York: John Wiley 1962.

MASON, B.: Meteorites. Am. Sci. **55**, No. 4, 429—455 (1967).

MATTAUCH, J.: Das Paar Rb^{87}-Sr^{87} und die Isobarenregel. Naturwissenschaften **25**, 189—191 (1937).

MATTAUCH, J.: Atomgewichtsbestimmung mit dem Massenspectrographen. Z. Anorg. Chem. **236**, 209—220 (1938).

MATTAUCH, J.: Stabile Isotope, ihre Messung und ihre Verwendung. Angew. Chem., A, No. 2, 37—42 (1947).

MAXWELL, D. T., HOWER, J.: High-grade diagenesis and low-grade metamorphism of illite in the Precambrian Belt series. Am. Mineral. **57**, 843—857 (1967).

MAYNE, K. I.: Stable isotope geochemistry and mass spectrometric analysis. In: SMALES, A. A., WAGER, R. L. (eds.): Methods in Geochemistry, pp. 148—195. New York: Interscience Publishers 1960.

McDOUGALL, I., COMPSTON, W., HAWKES, O. D.: Leakage of radiogenic argon and strontium from minerals in Proterozoic dolerites from British Guiana. Nature **198**, 564—567 (1963).

McDOUGALL, I., COMPSTON, W.: Strontium isotope composition and potassium-rubidium ratios in some rocks from Reunion and Rodriguez, Indian Ocean. Nature **207**, 252—253 (1965).

McDOUGALL, I., DUNN, P. R., COMPSTON, W., WEBB, A. W., RICHARDS, J. R., BOFINGER, V. M.: Isotopic age determinations on Precambrian rocks of the Carpentaria region, Northern Territory, Australia. J. Geol. Soc. Australia **12**, 67—90 (1965).

McDOWELL, C. A.: Mass Spectrometry. New York: McGraw-Hill 1963.

McELHINNY, M. W.: Rb-Sr and K-Ar age measurements on the Modipe Gabbro of Bechuanaland and South Africa. Earth Planet. Sci. Letters **1**, No. 6, 439—442 (1966).

McIntyre, G. A., BROOKS, C., COMPSTON, W., TUREK, A.: The statistical assessment of Rb-Sr isochrons. J. Geophys. Res. **71**, 5459—5468 (1966).

McMULLEN, C. C., FRITZE, K., TOMLINSON, R. H.: The half-life of rubidium-87. Can. J. Phys. **44**, 3033—3038 (1966).

MEYER, C. JR., AITKEN, F. K., BRETT, R., McKAY, D., MORRISON, D.: Rock fragments and glasses rich in K, REE, P in Apollo 12 soils: Their mineralogy and origin. Abstract. 1971 Lunar Sci. Conf., January 11—14, 1971, NASA, Houston, Texas, p. 245 (1971).

MITCHELL, R. H., CROCKET, J. H., McNUTT, R. H.: Isotopic composition of strontium in some South African kimberlites. Abstract, Trans. Am. Geophys. Union **50**, 346 (1969).

MITCHELL, R. H., CROCKET, J. H.: The isotopic composition of strontium in alkaline rocks of the Fen Complex. South Norway. Abstract. Symposium on volcanoes and their roots, Oxford. IAVEI, p. 137 (1969a).

MITCHELL, R. H., CROCKET, J. H.: The isotopic composition of strontium in some South African kimberlites. Contrib. Mineral. Petrol. 30, 277—290 (1971).

MOORBATH, S., HURLEY, P. M., FAURE, G., FAIRBAIRN, H. W., PINSON, W. H.: Evidence for the origin of mineralized Tertiary intrusives in southwestern states from strontium-isotope ratios. Abstract, J. Geophys. Res. 67, 3582 (1962).

MOORBATH, S., WALKER, G. P. L.: Strontium isotope investigation of igneous rocks from Iceland. Nature 207, 837—840 (1965).

MOORBATH, S., BELL, J. D.: Strontium isotope abundance studies and rubidium-strontium age determinations on Tertiary igneous rocks from the Isle of Skye, northwest Scotland. J. Petrol. 6, 37—66 (1965).

MOORBATH, S.: Evidence for the age of deposition of the Torridonian sediments of northwest Scotland. Scott. J. Geol. 5, Part 2, 154—170 (1969).

MURTHY, V. R., STUEBER, A. M.: Strontium isotopic composition of recent volcanic rocks from the Pacific Ocean Basin. Abstract. Trans. Am. Geophys. Union 44, 112 (1963).

MURTHY, V. R., COMPSTON, W.: Rb-Sr ages of chondrules and carbonaceous chondrites. J. Geophys. Res. 70, 5297—5307 (1965).

MURTHY, V. R., BEISER, E.: Strontium isotopes in ocean water and marine sediments. Geochim. Cosmochim. Acta 32, 1121—1126 (1968).

MURTHY, V. R., SCHMITT, R. A., REY, P.: Rubidium-strontium age and elemental and isotopic abundances of some trace elements in lunar samples. Science 167, 476—479 (1970).

MURTHY, V. R., EVENSEN, N. M., CIOSCIO, M. R., JR.: Distribution of K, Rb, Sr, and Ba and Rb-Sr isotopic reactions in Apollo 11 lunar samples. Proc. Apollo 11 Lunar Sci. Conf. 2, 1393—1406 (1970).

NICOLAYSEN, L. O.: Graphic interpretation of discordant age measurements of metamorphic rocks. Ann. N. Y. Acad. Sci. 91, Article 2, 198—206 (1971).

NIER, A. O.: Isotopic constitution of Sr, Ba, Bi, Tl, and Hg. Phys. Rev. 54, 275—278 (1938).

NOBLE, D. C., HEDGE, C. E.: $^{87}Sr/^{86}Sr$ variations within individual ash-flow sheets. U. S. Geol. Surv. Prof. Paper 650 C, C133—C139 (1969).

NOCKOLDS, S. R., MITCHELL, R. L.: The geochemistry of some Caledonian plutonic rocks; a study in the relationship between the major and trace elements of igneous rocks and their minerals. Trans. Roy. Soc. Edinburgh 61, 535—575 (1948).

NOCKOLDS, S. R., ALLEN, R.: The geochemistry of some igneous rock series; Part I, Calc-alkalic rocks. Geochim. Cosmochim. Acta 4, 105—142 (1953).

NOCKOLDS, S. R., ALLEN, R.: The geochemistry of some igneous rock series; Part II, Alkali igneous rock series. Geochim. Cosmochim. Acta 5, 245—285 (1954).

NOCKOLDS, S. R., ALLEN, R.: The geochemistry of some igneous rock series. Part III. Geochim. Cosmochim. Acta 9, 34 (1956).

NOLL, W. VON: Geochemie des Strontiums. Chemie der Erde, Ed. VIII, p. 507. 1934.

NORRISH, K., CHAPPELL, B. W.: X-ray fluorescence spectrography. In: ZUSSMAN, J. A. (ed.): Physical Methods in Determinitive Mineralogy. New York: Academic Press 1967.

NUNES, P. D., TILTON, G. R.: Uranium-lead ages of minerals from the Stillwater Complex and associated rocks, Montana. Geol. Soc. Am. Bull. **82**, No. 8, 2231—2250 (1971).

OBRADOVICH, J. D., PETERMAN, Z. E.: Geochronology of the Belt Series, Montana: Can. J. Earth Sci. **5**, 737—747 (1968).

ODUM, H. T.: Biogeochemical deposition of strontium: Inst. Marine Sci. **4**, No. 2, 38—114 (1957).

O'NEIL, J. R., HEDGE, C. E., JACKSON, E. D.: Isotopic investigations of xenoliths and host basalts from the Honolulu volcanic series. Earth Planet. Sci. Letters **8**, 253—257 (1970).

OVERSBY, V. M., GAST, P. W.: Isotopic composition of lead from oceanic islands. J. Geophys. Res. **75**, 2097—2114 (1970).

OZIMA, M., ZASHU, S., UENO, N.: K/Rb and (^{87}Sr/^{86}Sr) oratios of dredged submarine basalts. Earth Planet. Sic. Letters **10**, 239—245 (1971).

PANKHURST, R. J.: Strontium isotope and geochronological studies of the basic igneous province of northeast Scotland. Abstract. Ph. D. Thesis, Oxford Univ. 1968.

PANKHURST, R. J.: Strontium isotopic studies related to petrogenesis in the Caledonian basic igneous province of northeast Scotland. J. Petrol. **10**, 115—143 (1969).

PANKHURST, R. J.: The geochronology of the basic igneous complexes. Scott. J. Geol. **6**, 83—107 (1970).

PAPANASTASSIOU, D. A., WASSERBURG, G. J.: Initial strontium isotopic abundances and the resolution of small time differences in the formation of planetary objects. Earth Planet. Sci. Letters **5**, 361—376 (1969).

PAPANASTASSIOU, D. A., WASSERBURG, G. J., BURNETT, D. S.: Rb-Sr ages of lunar rocks from the Sea of Tranquillity. Earth Planet. Sci. Letters **8**, 1—9 (1970).

PAPANASTASSIOU, D. A.: The determination of small time differences in the formation of planetary objects. Unpublished thesis. Division of Geological and Planetary Sciences, California Institute of Technology 1970.

PAPANASTASSIOU, D. A., WASSERBURG, G. J.: Rb-Sr ages from the Ocean of Storms. Earth Planet. Sci. Letters **8**, 269—278 (1970).

PAPANASTASSIOU, D. A., WASSERBURG, G. J.: Lunar chronology and evolution from Rb-Sr studies of Apollo 11 and 12 samples. Earth Planet. Sci. Letters **11**, 37—62 (1971).

PATTERSON, C., TATSUMOTO, M.: The significance of lead isotopes in detrital feldspar with respect to chemical differentiation within the earth's mantle. Geochim. Cosmochim. Acta **28**, 1—22 (1964).

PATTERSON, C.: Characteristics of lead isotope evolution on a continental scale in the earth. In: CRAIG, H., MILLER, S. L., WASSERBURG, G. J., (eds.): Isotopic and Cosmic Chemistry. Amsterdam: North-Holland Publ. Co. 1964.

PAULING, L.: The Nature of the Chemical Bond. Ithaca, N. Y.: Cornell University Press 1960.

PETERMAN, Z. E., HEDGE, C. E., BRADDOCK, W. A.: Age of Precambrian events in the northeastern Front Range, Colorado. J. Geophys. Res. **73**, 2277—2296 (1968).

PETERMAN, Z. E.: Rb-Sr dating of Middle Precambrian metasedimentary rocks of Minnesota. Geol. Soc. Am. Bull. **77**, 1031—1044 (1966).

PETERMAN, Z. E., HEDGE, C., COLEMAN, R. G., SNAVELY, P. D., JR.: Sr^{87}/Sr^{86} ratios in some eugeosynclinal sedimentary rocks and their bearing on the origin of granitic magmas in orogenic belts. Earth Planet. Sci. Letters **2**, 433—439 (1967).

PETERMAN, Z. E., DOE, B. R., PROSTKA, H. J.: Lead and strontium isotopes in rocks of the Absaroka volcanic field, Wyoming. Contrib. Mineral. Petrol. **27**, 121—130 (1970).

PETERMAN, Z. E., HEDGE, C. E., TOURTELOT, H. A.: Isotopic composition of strontium in sea water throughout Phanerozoic time. Geochim. Cosmochim. Acta **34**, 105—120 (1970).

PETERMAN, Z. E., LOWDER, G. G., CARMICHAEL, I. S. E.: Sr^{87}/Sr^{86} ratios of the Talasea Series, New Britain, Territory of New Guinea. Geol. Soc. Am. Bull. **81**, 39—40 (1970).

PETERMAN, Z. E., CARMICHAEL, I. S. E., SMITH, A. L.: Strontium isotopes in Quaternary basalts of southeastern California. Earth Planet. Sci. Letters **7**, 381—384 (1970a).

PETERMAN, Z. E., CARMICHAEL, I. S. E., SMITH, A. L.: Sr^{87}/Sr^{86} ratios of Quaternary lavas of the Cascade Range, northern California. Geol. Soc. Am. Bull. **81**, 311—318 (1970b).

PETERMAN, Z. E., HEDGE, C. E.: Related strontium isotopic and chemical variations in oceanic basalts. Geol. Soc. Am. Bull. **82**, 493—500 (1971).

PIDGEON, R. T., COMPSTON, W.: The age and origin of the Cooma Granite and its associated metamorphic zones, New South Wales. J. Petrol. **6**, 193—222 (1965).

PILOT, J., RÖSLER, H. J.: Altersbestimmungen von Kalisalzmineralien. Naturwissenschaften **54**, 490 (1967).

PINSON, W. H., SCHNETZLER, C. C., BEISER, E., FAIRBAIRN, H. W., HURLEY, P. M.: Rb-Sr age of stony meteorites. Geochim. Cosmochim. Acta **29**, 455—466 (1965).

POLEVAYA, N. I., TITOV, N. E., BELYAEV, V. S., SPRINTSSON, V. D.: Application of the Ca method in the absolute age determination of sylvites. Geochemistry **8**, 897—906 (1958).

POLEVAYA, N. I., MURINA, G. A., KAZAKOV, G. A.: Utilization of glauconite in absolute dating. In: KULP, J. L. (ed.): Geochronology of Rock Systems. Ann. N. Y. Acad. Sci. **91**, 298—310 (1961).

POWELL, J. L., HURLEY, P. M., FAIRBAIRN, H. W.: Isotopic composition of strontium in carbonatites. Nature **196**, 1085—1086 (1962).

POWELL, J. L.: Isotopic composition of strontium in four carbonate vein-dikes. Am. Mineralogist **50**, 1921—1928 (1965a).

POWELL, J. L.: Low abundance of Sr^{87} in Ontario carbonatites. Am. Mineralogist **50**, 1075—1079 (1965b).

POWELL, J. L.: Isotopic composition of strontium in carbonate rocks from Keshya and Mkwisi, Zambia. Nature **206**, 288—289 (1965c).

POWELL, J. L., FAURE, G., HURLEY, P. M.: Strontium 87 abundance in a suite of Hawaiian volcanic rocks of varying silica content. J. Geophys. Res. **70**, 1509—1513 (1965).

POWELL, J. L.: Isotopic composition of strontium in carbonatites and kimberlites. Mineral. Soc. India, I. M. A. Volume, 58—66 (1966).

POWELL, J. L., DELONG, S. E.: Isotopic composition of strontium in volcanic rocks from Oahu. Science **153**, 1239—1242 (1966).

POWELL, J. L., HURLEY, P. M., FAIRBAIRN, H. W.: The strontium isotopic composition and origin of carbonatites. In: TUTTLE, O. F., GITTINS, J. (eds.): Carbonatites, pp. 365—378. London: Interscience Publishers 1966.

POWELL, J. L., SKINNER, W. R., WALKER, D.: Whole-rock Rb-Sr age of metasedimentary rocks below the Stillwater Complex, Montana. Geol. Soc. Am. Bull. **80**, 1605—1612 (1969).

Powell, J. L., Bell, K.: Recognition of contamination in igneous rocks using strontium isotopes. Geol. Soc. Am. Abstracts with Programs 1, Part 6, 37—38 (1969).

Powell, J. L., Bell, K.: Strontium isotopic studies of alkalic rocks. Localities from Australia, Spain, and the Western United States. Contrib. Mineral. Petrol. 27, 1—10 (1970).

Powell, J. L., Bell, K.: Isotopic composition of strontium in alkalic rocks. In: Sorensen, H. (ed.): The Alkaline Rocks. Wiley-Interscience (in press).

Pushkar, P.: Isotopic composition of strontium in Central American ignimbrites. Univ. Arizona Ann. Prog. Rep. No. C00-689-76 (1967).

Pushkar, P.: Strontium isotope ratios in volcanic rocks of three island arc areas. J. Geophys. Res. 73, 2701—2714 (1968).

Pushkar, P., McBirney, A. R.: The isotopic composition of strontium in Central American ignimbrites. Univ. Arizona Ann. Prog. Rep. No. C00-689-100 (1968).

Raguin, E.: Geology of Granite. Interscience: New York 1965.

Rankama, K., Sahama, T. G.: Geochemistry. Chicago: The University of Chicago Press 1950.

Read, H. H.: The Granite Controversy. London: Murby and Co. 1957.

Reesman, R. H.: Strontium isotopic compositions of gangue minerals from hydrothermal vein deposits. Econ. Geol. 63, 731—736 (1968).

Riley, G. H., Compston, W.: Theoretical and technical aspects of Rb-Sr geochronology. Geochim. Cosmochim. Acta 26, 1255—1281 (1962).

Ringwood, A. E.: On the chemical evolution and densities of the planets. Geochim. Cosmochim. Acta 15, 257—283 (1959).

Ringwood, A. E.: Genesis of chondritic meteorites. Rev. Geophys. 4, 113—175 (1966).

Roe, G. D.: Rubidium-strontium analyses of ultramafic rocks and the origin of peridotites. Twelfth Ann. Prog. Rep. M.I.T. 159—190 (1964).

Roe, G. D., Pinson, W. H. Jr., Hurley, P. M.: Rb-Sr evidence for the origin of peridotites. Abstract. Trans. Am. Geophys. Union 46, 186 (1965).

Roedder, E.: Liquid CO_2 inclusions in olivine-bearing nodules and phenocrysts from basalts. Am. Mineralogist 50, 1764—1782 (1965).

Sanz, H. G., Wasserburg, G. J.: Determination of an internal Rb^{87}-Sr^{87} isochron for the Olivenza chondrite. Earth Planet. Sci. Letters 6, 335—345 (1969).

Sanz, H. G., Burnett, D. S., Wasserburg, G. J.: A precise $^{87}Rb/^{87}Sr$ age and initial $^{87}Sr/^{86}Sr$ for the Colomera iron meteorite. Geochim. Cosmochim. Acta 34, 1227—1239 (1970).

Schreiner, G. D. L.: Comparison of the Rb^{87}-Sr^{87} ages of the Red Granite of the Bushveld Complex from measurements on the total rock and separated mineral fractions. Proc. Roy. Soc. London, Ser. A, 245, 112—117 (1958).

Schumacher, E.: Altersbestimmungen von Steinmeteoriten mit der Rb-Sr-Methode. Z. Naturforsch. 11a, 206—212 (1956).

Shaw, D. M., Moxham, R. L., Filby, R. H., Lapkowsky, W. W.: The petrology of some Grenville Skarns. Part II. Geochemistry. Can. Mineral. 7, 578—616 (1963).

Shields, R. M., Pinson, W. H., Jr., Hurley, P. M.: Rubidium-strontium analyses of the Bjurböle chondrite. J. Geophys. Res. 71, 2163—2167 (1966).

Shields, W. R., Garner, E. L., Hedge, C. E., Goldich, S. S.: Survey of Rb^{85}/Rb^{87} ratios in minerals. J. Geophys. Res. 68, 2331—2334 (1963).

Shima, M., Honda, M.: Determination of rubidium-strontium age of chondrites using their separated components. Earth Planet. Sci. Letters 2, 337—343 (1967).

SINHA, A. K., DAVIS, G. L.: Geochemistry of Franciscan volcanic and sedimentary rocks from California. Carnegie Inst. Wash. Yearbook 69, 394—400 (1971).

SMITH, A. G., HALLAM, A.: The fit of the southern continents. Nature 225, 139—144 (1970).

SORBY, H. C.: On the structure and origin of meteorites. Nature 15, 495—498 (1877).

SORENSEN, H. (editor): The Alkaline Rocks. New York: Wiley-Interscience. In Press.

STEIGER, R. H., HART, S. R.: The microcline-orthoclase transition within a contact aureole. Am. Mineral. 52, 87—116 (1967).

STEVENS, R. E., SCHALLER, W. T.: The rare alkalies in micas. Am. Mineral. 27, 525—537 (1942).

STEWART, F. H.: Marine evaporites, pp. Y1—152. In: FLEISCHER, M. (ed.): Data of Geochemistry. U. S. Geol. Surv. Prof. Pap. 440-Y (1963).

STRASSMAN, F., WALLING, E.: Die Abscheidung des reinen Strontium-Isotope 87 aus einem alten rubidium-haltigen Lepidolith und die Halbwertszeit des Rubidiums: Ber. Deut. Keram. Ges. 71, Abt. B, 1—9 (1938).

STUEBER, A. M., MURTHY, V. R.: Strontium isotope and alkali element abundances in ultramafic rocks. Abstract. Trans. Am. Geophys. Union 46, 186 (1965).

STUEBER, A. M., MURTHY, V. R.: Strontium isotope and alkali element abundances in ultramafic rocks. Geochim. Cosmochim. Acta 30, 1243—1259 (1966).

STUEBER, A. M.: Potassium, rubidium, and strontium in ultramafic rocks and minerals. Carnegie Inst. Wash. Yearbook 66, 42 (1967).

STUEBER, A. M.: Abundances of K, Rb, Sr, and Sr isotopes in ultramafic rocks and minerals from western North Carolina. Geochim. Cosmochim. Acta 33, 543—553 (1969).

SUMMERHAYES, C. P.: A geochronological and strontium isotope study of the Garabal Hill-Glen Fyne igneous complex, Scotland. Geol. Mag. 103, 153—165 (1966).

TATSUMOTO, M., HEDGE, C. E., ENGEL, A. E. J.: Potassium, rubidium, strontium, thorium, uranium and the ratio of strontium-87 to strontium-86 in oceanic tholeiitic basalt. Science 180, 886—888 (1965).

TATSUMOTO, M.: Isotopic composition of lead in volcanic rocks from Hawaii, Iwo Jima, and Japan. J. Geophys. Res. 71, 1721—1733 (1966).

TATSUMOTO, M.: Lead isotopes in volcanic rocks and possible ocean-floor thrusting beneath island arcs. Earth Planet. Sci. Letters 6, 369—376 (1969).

TAUBENECK, W. H.: An appraisal of some potassium-rubidium ratios in igneous rocks. J. Geophys. Res. 70, 475—478 (1965).

TAYLOR, H. P., EPSTEIN, S.: O^{18}/O^{16} ratios in rocks and coexisting minerals of the Skaergaard intrusion, East Greenland. J. Petrol. 4, 51—74 (1963).

TAYLOR, S. R., HEIER, K. S.: Alkali elements in potash feldspars from the preCambrian of southern Norway. Geochim. Cosmochim. Acta 13, 293—302 (1958).

TAYLOR, S. R.: The abundance of chemical elements in the continental crust — a new table. Geochim. Cosmochim. Acta 28, 1273—1285 (1964).

TAYLOR, S. R.: The application of trace element data to problems in petrology, pp. 133—214. In: AHRENS, L. A., PRESS, F., RUNCORN, S. K., UREY, C., (eds.): Physics and Chemistry of the Earth. Oxford: Pergamon Press Vol. 6. 1965.

THOMSON, J. J.: Emission of negative corpuscles by alkali metals. Phil. Mag. 10, 584—590 (1905).

TILTON, G. R., WETHERILL, G. W., DAVIS, G. L., HOPSON, C. A.: Ages of minerals from the Baltimore Gneiss near Baltimore, Maryland. Geol. Soc. Am. Bull. 69, 1469—1474 (1958).

TILTON, G. R., GAST, P. W., HEDGE, C. E.: Variation in the isotopic composition of lead and strontium from basalts. Abstract. Trans. Am. Geophys. Union 45, 109 (1964).

TILTON, G. R., DAVIS, G. L., HART, S. R., ALDRICH, L. T., STEIGER, R. H., GAST, P. W.: Geochronology and isotope geochemistry. Carnegie Inst. Wash. Yearbook 63, 250 (1964).

TOMLINSON, R. H., DAS GUPTA, A. K.: The use of isotope dilution in determination of geologic age of minerals. Can. J. Chem. 31, 909—914 (1953).

TUREKIAN, K. K., KULP, J. L.: The geochemistry of strontium. Geochim. Cosmochim. Acta 10, 245—296 (1956).

TUREKIAN, K. K., WEDEPOHL, K. H.: Distribution of the elements in some major units of the earth's crust. Geol. Soc. Am. Bull. 72, 175—192 (1961).

TUREKIAN, K. K.: The marine geochemistry of strontium. Geochim. Cosmochim. Acta 28, 1479—1496 (1964).

TURNER, F. J., VERHOOGEN, J.: Igneous and Metamorphic Petrology, 2d ed. New York: McGraw-Hill 1960.

TUTTLE, O. F., BOWEN, N. L.: Origin of granite in the light of experimental studies in the system $NaAlSi_3O_8—KAlSi_3O_8—SiO_2$ H_2O. Geol. Soc. Am. Mem. 74, 153 p. (1958).

TUTTLE, F., GITTINS, J. (eds.): Carbonatites. New York: Interscience. 1966.

UREY, H. C.: Diamonds, meteorites, and the origin of the solar system. Astrophys. J. 124, 623—637 (1956).

UREY, H. C.: Boundary conditions for theories of the origin of the solar system. In: Prog. Phys. Chem. Earth 2, 46—76 (1957).

UREY, H. C.: The early history of the solar system as indicated by the meteorites. Proc. Chem. Soc. 67—78 (1958).

VAN SCHMUS, W. R.: The geochronology of the Blind River — Bruce Mines area, Ontario, Canada. J. Geol. 73, 755—780 (1965).

VAN SCHMUS, W. R., WOOD, J. A.: A chemical-petrologic classification for the chondritic meteorites. Geochim. Cosmochim. Acta 31, 747—766 (1967).

VINOGRADOV, A. P.: The regularity of distribution of chemical elements in the earth's crust. Geochemistry 1, 1—43 (1956).

VINOGRADOV, A. P.: Average contents of chemical elements in the principal types of igneous rocks of the earth's crust. Geochemistry 7, 641—664 (1962).

VINOGRADOV, A. P.: Preliminary data on the lunar ground brought to Earth by automatic probe Luna-16. 1971 Lunar Sci. Conf., January 11—14, 1971. NASA, Houston, Texas 1971.

VLASOV, K. A. (ed.): Geochemistry and mineralogy of rare elements and genetic types of their deposits. Volume I. Geochemistry of rare elements. Translated by Z. LERMAN, Israel Program for Scientific Translations, Jerusalem, 1966 (1964a).

VLASOV, K. A. (ed.): Geochemistry and mineralogy of rare elements and genetic types of their deposits. Volume II. Mineralogy of rare elements. Translated by Z. LERMAN, Israel Program for Scientific Translations, Jerusalem, 1966 (1964b).

WANLESS, R. K., LOVERIDGE, W. D., MURSKY, G.: A geochronological study of the White Creek batholith, southeastern British Columbia. Can. J. Earth Sci. 5, 375—386 (1968).

WANLESS, R. K., LOVERIDGE, W. D., STEVENS, R. D.: Age determinations and isotopic abundance measurements on lunar samples. Science **167**, 479—480 (1970a).

WANLESS, R. K., LOVERIDGE, W. D., STEVENS, R. D.: Age determinations and isotopic abundance measurements of lunar samples (Apollo 11). Proc. Apollo 11 Lunar Sci. Conf. **2**, 1729—1740 (1970b).

WARDLAW, N. C.: Carnallite-sylvite relationships in the Middle Devonian Prairie Evaporite Formation, Saskatchewan. Geol. Soc. Am. Bull. **79**, 1273—1294 (1968).

WASSERBURG, G. J., ALBEE, A. L., LANPHERE, M. A.: Migration of radiogenic strontium during metamorphism. J. Geophys. Res. **69**, 4395—4401 (1964a).

WASSERBURG, G. J., WEN, T., ARONSON, J.: Strontium contamination in mineral analysis. Geochim. Cosmochim. Acta **28**, 407—410 (1964b).

WASSERBURG, G. J., BURNETT, D. S., FRONDEL, C.: Strontium-rubidium age of an iron meteorite. Science **150**, 1814—1817 (1965).

WASSERBURG, G. J., LANPHERE, M. A.: Age determinations in the Pre-cambrian of Arizona and Nevada. Geol. Soc. Am. Bull. **76**, 735—758 (1965).

WASSERBURG, G. J., BURNETT, D. S.: The status of isotopic age determinations on iron and stone meteorites. In: MILLMAN, P. M. (ed.): Meteorite Research. Dortrecht, Holland: Reidel Publ. Co. 1969.

WASSERBURG, G. J., PAPANASTASSIOU, D. A., NENOW, E. V., BAUMANN, C. A.: A programmable magnetic field mass spectrometer with on-line data processing. Rev. Sci. Instr. **40**, 288 (1969).

WASSERBURG, G. J., PAPANASTASSIOU, D. A., SANZ, H. G.: Initial strontium for a chondrite and the determination of a metamorphism or formation interval. Earth Planet. Sci. Letters **7**, 33—43 (1969).

WEBSTER, R. K., MORGAN, J. W., SMALES, A. A.: Some recent Harwell analytical work on geochronology. Am. Geophys. Union Trans. **38**, 543—545 (1957).

WEBSTER, R. K.: Mass spectrometric isotope dilution analysis. In: Methods in Geochemistry. New York-London: Interscience Publ. 1960.

WETHERILL, G. W., DAVIS, G. L., TILTON, G. R.: Age measurements on minerals from the Cutler Batholith, Cutler, Ontario. J. Geophys. Res. **65**, 2461—2466 (1960).

WETHERILL, G. W., BICKFORD, M. E.: Primary and metamorphic Rb-Sr chronology in Central Colorado. J. Geophys. Res. **70**, 4669—4686 (1965).

WETHERILL, G. W., TILTON, G. R., DAVIS, H. L., HART, S. R., HOPSON, C. A.: Age measurements in the Maryland Piedmont. J. Geophys. Res. **71**, 2139—2155 (1966).

WETHERILL, G. W., DAVIS, G. L., LEE-HU, C.: Rb-Sr measurements on whole rocks and separated minerals from the Baltimore Gneiss, Maryland. Geol. Soc. Am. Bull. **79**, 757—762 (1968).

WETHERILL, G. W.: Of time and the Moon. Science **173**, 383—392 (1971).

WHIPPLE, F. E.: Chondrules: suggestion concerning their origin. Science **153**, 54—56 (1966).

WHITE, A. J. R., COMPSTON, W., KLEEMAN, A. W.: The Palmer Granite — A study of a granite within a regional metamorphic environment. J. Petrol. **8**, 29—50 (1967).

WHITNEY, P. R., HURLEY, P. M.: The problem of inherited radiogenic strontium in sedimentary age determinations. Geochim. Cosmochim. Acta **28**, 425—436 (1964).

WICKMAN, F. E.: Isotope ratios: A clue to the age of certain marine sediments. J. Geol. **56**, 61—66 (1948).

WOOD, J. A.: Physics and chemistry of meteorites. In: MIDDLEHURST, B. M., KUIPER, G. P. (eds.): The Moon, Meteorites and Comets. University of Chicago Press 1963.

WOOD, J. A.: Meteorites and the origin of planets. New York: McGraw-Hill 1968.

WYLLIE, P. J.: Experimental studies of carbonatite problems: The origin and differentiation of carbonatite magmas. In: TUTTLE, O. F., GITTINS, J. (eds.): Carbonatites, pp. 311—352. London: Interscience Publ. 1966.

WYLLIE, P. J. (ed.): Ultramafic and related rocks. London-Sydney-New York: John Wiley 1967.

YODER, H. S., JR.: Calkalkalic andesites: Experimental data bearing on the origin of their assumed characteristics. Oregon Dept. Geol. Mineral. Ind. Bull. **65**, 77—89 (1969).

YORK, D.: Least-squares fitting of a straight line. Can. J. Phys. **44**, 1079—1086 (1966).

YORK, D.: The best isochron: Earth Planet. Sci. Letters **2**, 479—482 (1967).

YOUDEN, W. J.: Statistical methods for chemists. New York: John Wiley 1951.

Author Index

Subject Index

Minerals, Rocks and Inorganic Materials